Data Duped

Data Duped

How to Avoid Being Hoodwinked by Misinformation

Derek W. Gibson and Jeffrey D. Camm

ROWMAN & LITTLEFIELD
Lanham • Boulder • New York • London

Published by Rowman & Littlefield
An imprint of The Rowman & Littlefield Publishing Group, Inc.
4501 Forbes Boulevard, Suite 200, Lanham, Maryland 20706
www.rowman.com

86-90 Paul Street, London EC2A 4NE

British Library Cataloguing in Publication Information Available

Library of Congress Cataloging-in-Publication Data

Names: Gibson, Derek W., 1968– author. | Camm, Jeffrey D., 1958– author.
Title: Data duped : how to avoid being hoodwinked by misinformation / Derek
  W. Gibson and Jeffrey D. Camm.
Description: Lanham : Rowman & Littlefield, [2023] | Includes
  bibliographical references and index.
Identifiers: LCCN 2022056511 (print) | LCCN 2022056512 (ebook) | ISBN
  9781538179147 (cloth) | ISBN 9781538179154 (ebook)
Subjects: LCSH: Disinformation. | Internet—Social aspects. | Truth—Political
  aspects. | Information society—Political aspects. | Social media—Political
  aspects. | Mass media—Political aspects. | Truthfulness and falsehood—
  Political aspects.
Classification: LCC HM1231 .G53 2023 (print) | LCC HM1231 (ebook) | DDC
  303.3/75—dc23/eng/20221201
LC record available at https://lccn.loc.gov/2022056511
LC ebook record available at https://lccn.loc.gov/2022056512

∞™ The paper used in this publication meets the minimum requirements of
American National Standard for Information Sciences—Permanence of Paper
for Printed Library Materials, ANSI/NISO Z39.48-1992.

Contents

Prologue

No one wants to be deceived, especially when they are making important decisions. Misinformation and data can be devious and messy and entangled between good and bad intentions, which makes the task of finding the truth much more difficult. As data professionals, we work with data all the time, and yet we also find ourselves confounded by misinformation. There is no immunity to being data duped, although we hope the lessons ahead in this book will help boost your data superpowers and make you a data skeptic when needed and a data believer when appropriate. That said, we have spent endless hours finding the best examples and thoroughly researching every reference, fact, and number we present on this journey. By the end, some astute readers may find something we missed or new data that changes some of our examples. This is okay. We are not perfect and in fact, we encourage readers to seek out new information and scrutinize the data. This is among the best data dupe defenses. We invite readers to share updates and some of their own stories with us on our companion site: *www.DataDuped.org*. There you will also find more in-depth analysis of examples in this book, and a few new ones too.

We hope the pages that follow make an improvement in our reader's ability to navigate misinformation, make better decisions, and avoid being data duped.

Derek and Jeff

Who Are You?

> **Duper**. noun. a person who deceives or tricks someone
> **Dupe**. noun. a person who is easily deceived or fooled
>
> —Oxford Dictionary

Henry and Linda never really paid much attention to their retirement. They would put a little money aside when they could and mostly let Henry's 401(k) do all the work. Once a year, Henry's boss gathered all the employees together to review the company's benefits packages and select their annual enrollment in their health care plan, and maybe talk about pay raises, at least when they could. The latter always helped when there were bills to pay. Medical bills in particular—checkups, a few cuts requiring stitches—it was all part of raising a young family. Taking care of their kids was always a more pressing matter than their far-off retirement. They never really paid that much attention. Years rolled by and Henry did a little better than most. Their kids wanted for not. There were piano lessons, sports camps, and school field trips. They bought and sold cars. They sent their children to college. Retirement? Well, they never paid much attention to it.

Eddie was known to never put down his phone. And when he did, he was reading something else on his iPad. He was constantly reading the news, although mostly what was available on Twitter and Facebook. When he sees a news story stating "Stocks *Tumble* on Heels of Inflation Report: Dow *Plunges* 240 Points," he decides it's time to call his broker.

Meanwhile, Ted, a farmer in Iowa, is making plans for the future of his business. He needs a new tractor, a very expensive one, and cannot decide if the price he is being offered is fair and if he should take advantage of the manufacturer's zero percent financing. And why exactly was the bank quoting him 4 percent interest when there was clearly a better deal at 0 percent?

In central Maine, Teresa works seasonal jobs here and there when she is not busy helping her family during lobster trapping season. The pay is good but it does not last all year long and she is thinking about how much better her life would be if she were rich. She steps into the convenience store on Farmington Road—her usual stop on the way to work—and decides today might be her lucky day. She buys 10 state lottery tickets and smiles while thinking to herself, *You gotta be in it to win it.*

Each of these people may not know it, but they have been data duped.

***

Who are you? What drew you to this book and, more importantly, why do we think we can help you from being data duped? Clearly, you are a curious person. A curious person with an interest in data, facts, figures, and how things work. You don't need to be a "data professional" to benefit from this book, because data can be found everywhere. Any member of society, practically anywhere in the world, can benefit from a better understanding of how others use data to influence them. Data can be right in front of you, perhaps in a news headline, "Thousands Trapped in Flash Floods," or it can be as subtle as part of the algorithms that anticipate your needs and provide recommendations about the movies you like, or remembering you prefer the double-pump latte at your favorite coffee shop. The more aware you are of numbers in your life the more profound it seems that they are infiltrating nearly everything you do—and this is just the beginning of the data age! Think what it will be like one, two, or five years from now.

In the last decade, we have seen great advances in both the creation of data and the use of that data to influence through analytics. About 90 percent of all data the world has ever known was created in the last two years. About 2.5 quintillion bytes of data are created each day . . . and it's growing. So, no matter if you are a student, a farmer, a doctor, a banker, a business owner, or a data scientist, numbers are in your life one way or another. You will encounter data at work, at home, at dinner parties, at all sorts of social meetings—and of course online. People have a natural curiosity to "make sense" of information and a strong innate need to bring order or at least a bit of understanding to the numbers around them. Like the compulsive individual who cannot be in a room with a slanted painting on the wall without the need to straighten it, we all in a way feel a desire to bring order to disorder. When presented with numbers, especially an overwhelming amount of numbers, it can be discomforting. How are we supposed to use all these numbers to make any sort of decision? Do the numbers support the decision you wanted to make before you saw them? How do you best use the numbers to make sure you are not making a mistake? In other words, we have a need to sort the numbers, put them into groups, seek patterns, and perhaps find confirmation and comfort that subdues the disruption we may have felt when we first encountered them. In short, we are humans and we are all vulnerable to being confused or misled by too much information.

In this book, we will look at all types of decisions in your life from the everyday "common" decisions such as shopping and driving directions, to larger "life" decisions like buying a home, planning for college, and saving for retirement. Being data duped unintentionally by the numbers can be costly. The benefit of being a little versed in the usage of numbers is to be proactive and hopefully make better decisions. Using a combination of skepticism, an understanding of probability, and some math skills can be revealing and useful.

We will also explore how companies can be data duped. Businesses big and small need to make decisions all the time, and hopefully these are based on data. If not, perhaps the first lesson in *Data Duped* is "Show me the numbers." Companies making decisions without data, or improperly using the data, are at risk of falling behind their competition. As companies grow in size and complexity, they are also growing in the amount of data they are collecting, often in disparate ways. This is likely because their data infrastructures are not created with analytics in mind—they are gathering information from new sources, new methods such as natural language processing, and new platforms (i.e., mobile, IoT, third party, etc.) and do not have the capacity to bring it in as fast as it is being created. With a disarray of data, it is easy to see how quick analysis of the data can be misleading. Being prepared to challenge findings and ask appropriate questions can improve results.

Being misled by data is not a new thing. From the 1880s Ponzi schemes, to Alexander the Great's invasion across the Alps, data have been at the heart of decisions big and small. What is new is the amount of data we, as regular everyday people, encounter. Everything from your smartphone to your refrigerator is creating data and presumably using it in ways that are either helpful to you or the companies creating these devices. We will even suggest there are companies gathering data from their new products that they do not yet know how to use. Data are easily created, while determining their usefulness is a more difficult task. This may lead to forced analysis to justify the existence of the data and worse—a host of unnecessary decisions based on data without a purpose. Data by itself is not misleading, but used without good curation, it can lead to misleading conclusions and bad decisions.

Data analytics when used fairly and with good intentions can improve decision-making. However, at the twilight of the analytics era, there are still many unknowns—is facial recognition good or bad? Is there bias in its training data that will perpetuate bad behavior? Can the past adequately predict the future? All of this may or may not be true. What is most important is how we prepare ourselves for our encounters with data, how we can be more informed, how we can ask the right questions, and ultimately how we can avoid being data duped.

THE FACES OF DATA DUPED

Just like any story of fiction or real life, there are protagonists and antagonists. In *Data Duped* there are antagonists, and their role is using data against the truth, transcending transparency to further a decision, an idea, or belief. We see this broadly in everyday personal decisions, decisions at work, the news, politics, and religion.

THE DATA DUPERS

The organizations and people misusing data, the data dupers, can be placed into three categories:

- *Unwitting Data Duper*—an individual who may not be purposely deceiving their audience. The dupe is a result of bad data, outliers, or poor curation of data.
- *Devious Data Duper*—the so-called deliberate dupers. They are motivated by their ideas and they want to influence and manipulate others. They might be guilty of handpicking the data and perhaps some confirmation bias, which we will explore further. They filter the data for what they need and leave out the rest.
- *Fallible Data Duper*—a result of humans' poor ability to deal with big numbers and accurately make estimates and, most importantly, our memory. Yes, our memory and ability to accurately recall information and synthesize it into a conclusion leaves us vulnerable. In academic papers, this process of using prior studies to build conclusions is called a meta-analysis. There is a certain framework and methodology to doing this right, and it almost always concludes with the need to do further research. However, most people are not well-trained data analysts. They are folks like you reading this book out of pure curiosity. Curious and skeptical people are the perfect audience for *Data Duped* because they can both benefit from our narrative and perhaps recognize they, too, are also vulnerable to a data duper.

  The fallible data duper pieces together sometimes disparate but similar ideas they have "*heard here and there*." Much of it is likely true, credible, and related to their particular point of view. However, *how* they remember those facts is neither consistent nor without bias. The fallible data duper is not purposely filtering the data like the devious data duper. It is more a function of how the human mind works when recalling facts. It is how your uncle Dilbert will mix together real facts and opinions. How the combination of multiple sources can lead him to a point of view all the while based on data. He is not a devious data duper and not a victim of bad data. He believes the data, but that does not mean you should too.

The protagonists of this journey are the receivers of misleading data. They are in two broad categories—the believers and unconfident skeptics.

THE DATA DUPES

- *Believers*—They accept most things they see and read as true. Numbers are not their foray, and in fact numbers may make them uncomfortable. They may be the ones most likely to react to online social media posts.
- *Unconfident Skeptics*—They are unsure of the data, but also unsure of knowing where to begin to seek the truth. What are the questions someone should ask when presented with data? What is an outlier and how does it influence averages?

JUST "ABOUT" RIGHT

With our cast of characters set, let's walk through a story of a potential data dupe that begins right here with you. Earlier we wrote "About 90 percent of all data the world has ever known was created in the last two years." Really? That sounds amazing. Our intention was not to deceive, only to quote a highly recognized reference about the growth of data and explosion of information. This claim is referenced so often in presentations and articles it has become folklore. A Google search for this statement returns millions of results, and yet we should wonder about this claim.

What should we make of the fact that the claim uses the wording "about 90 percent"? Why is it not more precise? Is the ambiguity intentional because it is a big number or because the writer was uncertain? And with ambiguity, does the statement become more proactive and stunning? It seems factual—a statement placed at the right point between common knowledge and unbelievable. Lean too far in one direction or the other and the headline fails to captivate. Sometimes "roughly" and "about" are an indication that something needs investigation.

What other questions might help us decide if this claim has merit? If the intention of the original headline was to gain a reaction, we might need to know more about who wrote it and why.

The claim appears to have come from an IBM study around 2012. However, the original source remains ambiguous even to a Google search. Some articles attribute the quote to an article by Åse Dragland in *Science Daily* in 2013.[1] Although, most quote it without reference to *any* source and that piques our interest.

Could "*About* 90 percent of all data . . . be created in the last two years"? How could anything fit this definition—cars on the road, websites, cells phones manufactured? It is technically possible, and using a spreadsheet tool like Excel there are some ways to estimate the growth of something that fits, but this usually requires growth from near zero, and data has been around for a long time.

So, we've asked some questions. Based on this path of investigation, it seems that the statement is plausible for 2012, but no longer true. It does seem likely that the amount of data we generate in a year continues to grow, maybe even at an exponential rate.

The claim about the amount of data also gets associated with another similar and widely quoted one from former Google CEO Eric Schmidt in 2010. At Google's 2010 Atmosphere convention, Schmidt made this statement: "There were five exabytes of information created between the dawn of civilization through 2003, but that much information is now created every two days." The *dawn of civilization* is similarly provocative and an exabyte is a measure of a very, very large amount of data. It was stunning and we can image members of the audience being in awe. So stunning that later it was learned that the five exabytes of data was generated in the *year* 2002. Not every two days. Not from the beginning of time through 2002, just in one year.[2] Generating a year's worth of data every two days is still impressive growth, but not quite as impressive as since the *dawn of civilization*. This slight misstep in referencing the data serves as a good example of how even very astute data-aware people can be duped by the numbers. Was Schmidt just in a hurry or were the numbers too believable that he did not question them? Did he default to believing when he should have instead performed some quick mental math, questioned the source, and considered other possibilities? Of course, he should have done this, but the numbers lulled him into easy acceptance. He was speaking about the *dawn of civilization*. He had reference points. He had numbers. And, he was data duped.

***

In this book, we hope to enhance your ability to be a healthy skeptic of claims made with data. Many, like the one we used earlier in this chapter, are designed and used to catch your attention. Our goal is to help you recognize when a claim might need further investigation and how to ask the right questions. Given that verifying or debunking claims takes time, you will need to decide when the extra effort and how much extra effort is valuable to you. Going forward, we will share a lot of other examples, financial, health-related, political, and others that might have real consequences for important decisions you will make.

At the opening, we introduced you to Henry, Eddie, Ted, and Teresa and their encounters with numbers. Henry's retirement planning tools provided by his employer relied on simple "straight-line" assumptions about how their money would grow over time (i.e., 6 percent per year *every* year). It is simple and unfortunately unrealistic. Eddie is overreacting to sensational headlines without context—stock market changes of 240 points seems dramatic; however, as a percentage change it is small. The context is missing from the headlines as we often see, and some readers are unprepared for how to interpret the information.

Ted is making a big financial decision and has conflicting information about the cost of getting a loan. The data deception is the manufacturer is selling the tractor at a higher price than needed to compensate for giving a *free* 0-percent interest loan. Finally, Teresa sees the lottery as a good bet, when it is not. Most likely she will lose the money she spent on lottery tickets. The details are in the fine print, however the marketing behind lottery tickets overemphasizes the possibility of winning. She also believes there is no way to win without buying a ticket, while there have actually been a few examples where people *do* win without buying a ticket. We will dive into the numbers and show you how all these examples are possible.

2

The Data Age and You

THE HISTORY OF DATA SCIENCE AND ANALYTICS

Many know the name Florence Nightingale. You know her fame as an English caregiver for establishing nursing techniques and saving lives during the 1850s Crimean War. She legitimized the role of nurses and vaulted rightly to become a heroic figure. She challenged and pushed boundaries. She was known as the "lamp lady" for her relentless care of her patients. She often checked on them throughout the night hours, carrying her lamp to guide her way from patient to patient, a sign of not only a caring person, but one who persevered, overcame obstacles, and paid attention to the details. What you may not know is Florence Nightingale was also a data scientist.

The term "data scientist" is somewhat new. It was barely a real term before 2008. By the time Tom Davenport and D. J. Patil described data scientists as the "sexiest job of the 21st century" in 2012, it was a role many had begun to hear about. A data scientist can blend data, statistics, and scientific methods toward insights. The insights rely on the curation of knowledge (data), technical tools to provide some summarization, and, perhaps most importantly, context for the problem being solved. A data scientist, in other words, is the trifecta of analytics and lies at the intersection of someone who has data, knows enough about what to do with it, and knows why it is important to do so.

In this manner, Florence Nightingale was a data scientist, although she came about her knowledge and context in an unconventional way. In the mid-1880s as a daughter of a wealthy English landowner, she was not expected to get a formal education and pursue a career. She did not attend university but rather was schooled by her father, who acquiesced to his daughter's curiosity. He taught history, philosophy, and, of course, math and statistics. She would later receive training in medical care, but it was her interest both in medicine and math that led to her accomplishments.

The Crimean War was terrible in ways most wars are not. There certainly were clashes of armies, loss of life, and injuries on the battlefield, but what made this war more horrible was the number of people who died *after* the fights. A staggering 80 percent of deaths from this two-year war came from disease. The diseases were not all related to battle wounds, but other things like dysentery, cholera, typhus, and typhoid. The war between the Russian Empire and the Ottoman Empire, with the latter supported by Britain and France, was particularly hard on the Russians, who experienced 10 times the number of deaths from disease rather than fighting. The British had more than six times the number of deaths from disease—more than 17,000. This is where Nightingale first began her endeavor with numbers.

Upon her return to Britain, she published a book called *Notes on Matters Affecting the Health, Efficiency, and Hospital Administration of the British Army*, a rather long title that is often referred to simply as *Notes on Nursing*. It became a standard for nursing education in the Nightingale training school she began with St. Thomas's Hospital in London. It is filled with data, charts, and tables with her narrative and interpretation. There are statistics and ratios comparing the survival of foot soldiers to cavalry and measurements of cubic feet of hospital space per man along with ratios of food and supplies. The data and its detail are extensive, but she describes what we would call data science today. She started with the data. She filtered and sifted numbers looking for patterns and root causes of disease and death. She applied her medical knowledge. She found insights. She noted patterns and correlations and pondered how the data could inform future outcomes. She made recommendations. Recommendations such as handwashing and hospital ventilation that would later save lives and dramatically reduce the number of preventable disease-related deaths. She was aware in her hundreds of pages of details that she might not be able to adequately explain her findings with numbers alone. Numbers in charts and tables sometimes can go only so far to convince others to make decisions. She overcame this by creating remarkable data visualizations—making her point with a picture rather than numbers alone. Her creativity in telling a data story visually resulted in a picture shaped like a pie, where each slice represented a month in the year of the war. Within each slice, three subsections radiated from the center depicting the number of deaths from battle, other causes, and disease, respectively. Rotating clockwise around the picture, viewers can see how the war unfolded each month. At first, the number of deaths was small, but as the war continued the increased fatalities showed a dramatic picture that significantly more were a result of disease than any other causes. Her data and its depictions in "Nightingale Graphs" not only moved forward her ideas, they also shaped generations of health care workers and data scientists alike.

DRIVING THE DATA AND ANALYTICS ERA

Faster and Cheaper Computing Power

Early applications of data to provide understanding and solve problems evolved from the days of Florence Nightingale and the Industrial Revolution to the Age of Industry. A period of rapid scientific discovery and factory-based mass production created an increasing demand for applying math to work. There was also a need to do it more efficiently than the manual efforts of people in a room with sharp pencils running the numbers the old-fashion way. Indeed, there was a demand for powerful automation and computers.

In 1944, International Business Machines, now known more commonly as just IBM, introduced their first "Main Frame" calculating machine. The Automatic Sequence Controlled Calculator (ASCC) could perform three additions per second and one multiplication every six seconds. A few years later that performance increased to thousands of calculations per second.[1] The ASCC was a massive machine measuring 51 feet long and 8 feet tall and weighing 5 tons.

The early mainframe computers worked much like their current-day decedents relying on logic switches. These switches were either on or off and in this manner indicate a one or zero, which is the basis for how all computers store data and make calculations. Early computers used mechanical relays for logic switches but soon were replaced by vacuum tubes. Then in 1947, a significant event happened when Bell

Lab engineers John Bardeen, Walter Brattain, and William Shockley developed the first working transistor. A transistor is also like a relay switch that turns on or off when given an electronic signal. By the time transistors were being mass-produced in the early 1950s, they were dramatically smaller than traditional vacuum tubes or relays, were much more reliable, and were by far cheaper to produce. A typical transistor that popularized transistor-AM radios in the 1950s and 1960s was about only four millimeters. You could easily fit 50 in the palm of your hand. The invention of transistors was truly a milestone for a broad category of electronics, including computers, which relied on these little switches to process data.

War and the space race further spurred innovation in computer technology. By the time of the Korean War in 1952, IBM introduced the IBM 700 series, becoming one of the first large-scale mass-produced computers that were so in demand by companies around the world. In the early 1960s, IBM mobile mainframes such as the IBM 1401 were packed onto army trailers and deployed to the field to assist as part of command centers.[2] The Apollo space mission also developed smaller computers for their spacecraft guidance and navigation systems that eventually transported astronauts to the moon.

When we need computers to do more things and run faster, then we need more transistors working together. The number of transistors impacts the capability of computers. At first, transistors were assembled on circuit boards and soon they were grouped onto integrated computer chips (IC). The first silicon-based computer chips started to emerge in the 1960s with the first commercially produced chip, the Intel 4004, hitting the market in 1971. From that point forward there has been an ever-increasing exponential growth in the number of transistors incorporated into computer processing units (CPUs), which are the central calculators of computers. The 1971 Intel 4004 chip had 2,250 transistors. By 2000, Intel's Pentium 4 chip boasted 42 *million* transistors. The most recent offering from AMD, the AMD Epyc Rome chip, has more than 39 *billion* transistors etched on its chip.

It is important to note that while automating calculations in computers was boosted exponentially by adding more transistors over time, there was also a dramatic reduction in costs. The cost per million transistors on a computer chip decreased from $5.27 in 2005 to a mere 23 cents today. And that Intel chip from 1971? Although it individually cost $60 ($400 adjusted to current dollars), it would take more than 400 of them and cost more than $100,000 for the equivalent of 1 million transistors. The cost barriers to computing have become an enabler to how we use data.

Fast-Forward to Cloud Computing

Taking advanced computer processing to the next level are "supercomputers" and cloud computing. "Cloud computing" refers to doing work on other computers. In a way, it is a return to the hub-and-spoke model of the early mainframe computer days, where your local computer is simply a terminal to the main central computer. Cloud computing offers relatively inexpensive access to large multiprocessor computers that are scalable to your needs. Providers such as Google Cloud, Microsoft Azure, and Amazon Web Services are the current leaders in this new offering. It can be ideal for smaller and mid-sized companies that could not before afford to invest in such technology, enabling machine learning and large-scale data analytics. Suddenly companies that would not be exploring their data can apply analytics to their business.

A Recipe for Change: The Analytics Era

Throughout history, there have been times when innovation, particularly in manufacturing, rapidly changed how we do things. The discovery of fire and the development of the Iron Age are classic examples of revolutionary change. We are entering a new period that is transforming nearly every company into a technology company, where individuals are impacted (hopefully in a positive way) by the development of data and its application. This era is sometimes referred to as the Fourth Industrial Revolution, but linking it solely to industrialization, which conjures up images of manufacturing, understates its scope. A better term is the analytics era.

The First Industrial Revolution spanned between 1765 and 1840. It truly was a revolutionary time for manufacturing with developments in machinery for specialized tasks such as milling machines. Steam power and harnessing into engines that propelled factories to develop machinery that relied less on human power and more on invention had

tremendous impacts on output. The textile industry was radically transformed, with factory output improving 500 times. The development
of the cotton gin that separated raw cotton from its seeds, improved
the processing time by a magnitude of 50. Iron-making even evolved
through the adoption of new processes using different heating methods
yielding greater quantities and furthering the building of machinery. It
was time that transformed the manufacture of clothing from a hand-
built task to a machine-built task, shifting jobs and benefiting consumers with more choices and lower costs.

The Second Industrial Revolution beginning in the 1870s saw the
development of mass production. It introduced an era of the assembly
line and notably the Ford Model T—one of the first mass-produced
cars. The assembly line broke the work task into individual steps where
one worker repeatedly performed the same task over and over. This
transformed manufacturing from a craftsman's process, where one
worker completed the assembly of a whole product before moving on
to the next, to a task-focused process. The improvements were so great
that the Model T assembly line could produce a car every two minutes.
In fact, one of the bottlenecks of early production was waiting for paint
to dry. The only paint that dried quickly enough was black, which is
the reason all Model Ts built before 1926 were black. Assembly line
production also decreased the cost of the car from $825 in 1908 to $575
in 1913.

The Third Industrial Revolution was sparked by the integration of
electricity into manufacturing and the addition of electronics spurred by
the invention of the transistor and microprocessors. Beginning in 1969,
factories saw more integration with electronically controlled machinery
(robotics) and control processes. Electronic products such as radios and
televisions were miniaturized, made more quickly, and less expensive
for consumers.

As each of these revolutions demonstrates, the introduction of a new
process or technology can have a cascading impact on industry. There
are new jobs requiring new skills, while other jobs fade away for lack of
necessity. There are also incredible benefits for consumers, often resulting in more affordable products or the introduction of entirely new ones
that did not exist before.

The analytics era shares these characteristics as well. Like the invention of the steam engine, the automation of the weaving looms, and

the creation of Ford's assembly line, this revolution required a few things to make it happen. As we have described above, there have been improvements in computing power—the capability of packing more and more transistors onto a silicon chip, all the while decreasing the costs to use it. There have been improvements in the algorithms used to solve problems such as optimization methods. Improvements in teaching computers tasks through efficient coding, nudged along by free open-source computer languages, created new methodologies accessible to big and small companies alike. The 1990s and 2000s saw more online commerce with the increasing capture of data from consumers and their transactions—even before most companies knew what to do with it.

As computer technology became less and less expensive, there was, and continues to be, more integration of computers into everything. Suddenly it seems everything is "smart." There are our smartphones, smart watches, smart appliances, smart light bulbs, smart cars, and more. The typical modern car collects 25 gigabits of data per hour— information on how fast (or slow) we drive, the performance of the engine, and even how much we weigh. If you have a new washing machine, it is collecting data, too, and knows how many times you have washed your clothes. The wide range of connected devices spawned the term "Internet of Things," or simply IoT. If there is something generating data, you can be assured someone has figured out how to connect it to a computer. One use case tracked honey bees by attaching small sensors to the bees, thereby allowing researchers to gather data about their daily flights to and from the hive.

In the data science circles, we talk a lot about how data are more in the hands of the people than ever before. Your data, as well as the data of others (anonymously), are available for wide consumption, for application to solve problems big and small, from the best marketing slogans to the most hopeful methods to cure disease. What is sometimes absent in that conversation about the democratization of data is the democratization of computing power and access to technology platforms. In the past, data and computing power were more expensive, and less accessible, except only to the wealthiest of companies and the best universities and colleges with dedicated computer science departments. Now nearly anyone with the most basic computer can access data and analytic power for almost no cost. This has leveled the playing field, so to speak, allowing many more people to participate. And where there are more participants, there is more investigation of its uses and more innovation.

All of these factors—less expensive and more powerful computing, better programming methods and technologies, and an abundance of data—soon collided in the mid-2000s. Like that scene in the movie *Back to the Future* where a well-timed lightning bolt strikes atop the clock tower, transporting Marty McFly through time, the analytics era had begun as a result of an amazing combination of powerful events. We are only a few years along this transformation and many years removed from its lasting effects.

Do you need more evidence the analytics era is transformational? Take a look at the most recent ranking of the largest companies in America based on their overall value. The top five companies are all technology companies (Microsoft, Apple, Amazon, Google, and Facebook [Meta]). Just ten years ago, only two of these were in the top ten (Microsoft and Apple) and the others were Exxon, Walmart, Berkshire, General Electric, and Proctor and Gamble. The analytics era has permanently shifted the economics of corporate America. What is more striking is how dominant these companies are compared to the rest of the top twenty companies. They are in a league of their own. Amazon, which reinvented the retail model created by Walmart, is now more than three times Walmart's size. Meanwhile, Microsoft is about the same size as Johnson & Johnson, Walmart, Visa, and JP Morgan combined.

Going back to 2005 you would barely see any data companies in the top 20. And the technology companies you did see, like Intel, were mostly makers of hardware, not yet embracing data and analytics as their core business. The list in 2005, similar to 2010, was dominated by Exxon and General Electric. In 2005, it was about energy, banking, manufacturing, and things you could buy. The current ranking shows how much that has changed. Amazon does not make things; they use data to know what things to make, sometimes in a highly customized manner. The power of these companies is not the physical products, rather it is the data they curate, control, and analyze. Unlike previous revolutions, which gave companies and nations economic advantage through the ability to manufacture things more efficiently, build new methods, and harness energy, the new era is all about data.

HOW DATA ARE INFILTRATING OUR LIVES

People will ask how will data really impact our lives and why is a book about avoiding being duped necessary? Many believe the data, the real deceptive stuff, is far off and away from their everyday lives, trapped in a computer, trying to generate Bitcoins, managing a telecom server, or working to solve some obscure academic pursuit. They believe data are not directly involved in their lives, but it is. It is everywhere.

Take a drive in your car to the grocery store and try to spot the places along your journey where data are collected, analyzed, and influencing you. Perhaps you do not see it, and this is natural. Data and its applications are sometimes quite subtle.

As you leave your home in your car, you likely have already logged into Google or Apple maps and picked your destination. Travel there often and Google will have already suggested your destination. Once the endpoint is confirmed, Google runs some calculations using data from your phone's location, its movement, the traffic along various possible routes, construction, and more to recommend the best path to follow. Every few seconds it will send a little bit of data about how fast your car is moving back to Google to help update their traffic predictions. The process cascades to other drivers around you as it also updates their routes. Too many cars moving too slowly and all of a sudden, the route on your map turns red. When it slows more, then the process repeats to find you the best alternative route.

The stoplight ahead turning red is not a random event either. The data from your phone and all the other drivers who have traveled through this intersection recently are helping predict traffic flows—the times of day when there are more or less cars. Using this information, the computers controlling the lights are adjusting the timing cycles to optimize the flow of cars through each intersection. Optimizing the timing of lights can get more cars through busy parts of town, solving a problem that otherwise would have been solved by building a wider road. Imagine data being more effective than building a whole road!

In Bellevue, Washington, analysts are taking their traffic data one step further . . . into the future. Using data from traffic cameras, they are applying artificial intelligence (AI) to recognize objects and search for patterns like collisions. In the old method of traffic analysis research, they would have to rely on traffic accident data, and after some time when one intersection or another showed more accidents, then an on-site study would be conducted. Adjustments could be made to traffic lanes, the timing of lights, and signage to avoid additional accidents. Bellevue's new approach seeks to avoid the accidents in the first place and rather focused on near collisions. Their computers could "see" the participants—cars, bicycles, motorcycles, and pedestrians—and could calculate speeds and near collisions, not an actual accident but an oh-my-God-I can't-believe-we-just-survived event. The technology analyzed 40 intersections, had 8.5 million participants, and witnessed more than 20,000 near-collision events. And then it did what all great computer systems do, it *learned*. The information from the patterns it saw allowed analysts to make adjustments and reduced actual future accidents by 60 percent.

Assuming you have driven your route safely, avoiding accidents and traffic alike, you arrive at the grocery store. When you enter the store, infrared sensors in systems like Kroger's QueVision may be counting you as you walk in. You are also counted when you are at the checkout line to ensure your wait time is held to a minimum. Your cell phone may connect unknowingly to the Wi-Fi beacons, allowing anonymous tracking of movement about the aisles. The data analysts will know how long you were in the produce section and if you paused to look at certain display items. Using your rewards card at checkout? Then they will also know what was in your shopping cart, and if you have provided information when you registered your rewards card, then they may know a few more things about you, such as where you live, your age, your estimated income, and your marital status.

As you leave the store, perhaps by way of the cashier-less checkout, your transactional data, the data list of the items in your cart, are all whisked away to the data cloud and added to your digital profile.

Your short journey to the store was filled with many "data moments"—both the collection of data and the application of it to (hopefully) improve your experience. The placement of items in the store and their arrangements on the shelves were not random, but a result of your data and that of all the other shoppers like you. In a way, your data were used to nudge you along either a particular road or in a shopping aisle toward a particular end. As we mentioned, data can be subtle in real life and ripe for data duping along the way.

DATA AND THE FUTURE OF WORK

A lot is being written about how the analytics era is and will continue to change how and what we do at work. The topic is commonly known as the "Future of Work." We are not sure that is the most creative title but at least it is descriptive! Most of the research points toward how the workplace will be transformed with the ongoing integration of AI assistants, recommendation engines, and automation processes, including robots, to replace routine tasks. This is a result of the analytics era. As a consumer, beyond the obvious transitions to online ordering, we have seen more "people jobs" move to other platforms. For example, when you check in at the airport, there are fewer gate agents and more self-serve kiosks to check in and even tag your luggage. If you have been to a McDonald's restaurant lately you may have noticed more locations are using self-ordering kiosks. These will route you through your order swiftly and routinely ask "do you want fries with that?" Eventually these will be able to recognize you if you are a regular customer and customize the order suggestions they offer. This is most likely how you will encounter data in your everyday lives.

Since this chapter is about data and you, it is worth noting how data and its use are evolving in a rapidly changing marketplace. Needless to say, there will be more data now than ever before. Later in this book, we will cover topics specific to data and decisions in the workplace. For now, it is important to recognize how data are seeping into the marketplace, so you are aware of when and where data are being used to guide you rather than you unknowingly being misguided.

As a consumer and recipient of data and analytics, you hopefully will enjoy the benefits. However, in the short term, your expectations might need to be a bit tempered. Here is how MIT professor Maria Yang describes it: "There's this romantic notion that everything is going to be automatic. The reality is you're going to have tools that will work with people and help make their daily life a bit easier."

What we are seeing are more subtle changes as data and analytics integrate with our workplaces. As Maria Yang suggests, it is not like the world is suddenly going to be taken over by robots (we will continue to do our best throughout this book to remind readers that data analytics is not like the movie *Terminator* and all those other dystopian movies).

We see analytics working alongside people in most jobs that are not repetitive, helping to augment their expertise.

Take the example of IBM Watson, which is IBM's branded product for collaborative AI. IBM Watson has been tailored for medicine and is helping doctors with patient treatment decisions. In one study doctors use information about a person's cancer cells to develop tailored therapies using Watson to help with the task. Watson helped them discover and change treatments that the doctors did not initially consider in about 13 percent of the cases reviewed. The primary benefit was Watson provided new information from clinical studies to inform the doctors on the decision board, leading to the treatment change. One reason for Watson's ability to provide new information is it could *read*. In addition to medical databases and detailed information about the patient, Watson can read text information and understand its meaning using a data science process called natural language processing (NLP). Using NLP Watson can access medical study publications through Clinical-Trials.gov and become the most well-read member of the medical team. According to ClinicalTrials.gov, there are more than 600 published studies *each week*. That's an amazing number that works out to more than 30,000 studies each year! Of course, a medical doctor only needs to focus on those relevant to their specialty, which may only be a few thousand papers to read. Having Watson read all those papers is like having your Alexa skim the morning newspapers for you and share only the topics that are most related to your day. A mere convenience for you, and potentially a life-saver for the medical community.

There are many more practical use cases like IBM Watson being deployed and studied in a variety of workplace situations from your investment advisor, known as robo-advisors, your auto mechanic, and even the people making hiring decisions at big companies. All of these and others are embarking into new territory using data.

PRIVACY, DATA, AND YOU

There is no doubt that privacy remains a top concern for people as more companies become data collectors, consumers, and purveyors. When we mention data deception and the misuse of data, we sometimes jolt the worst of fears in people, and some of that may not be unfounded.

At the beginning of the telephone industry, individual switchboard operators were needed to connect one caller with another. At the time a caller would "ring up" the operator and ask them to connect the call with whom they wanted to speak. The operator completed this task by plugging cables—one for the original caller and the other for the recipient—into a switchboard. (Now, perhaps that 1970s song "Operator" by Jim Croce finally makes sense.) Electronic switching to connect calls did not come until later, making the job of a switchboard operator popular in the 1950s and 1960s. One feature of this method of connecting people was switchboard operators had the ability to listen in on conversations. It was merely intended to connect the two callers and then the operator would leave the conversation. Listening to calls was strictly forbidden due to privacy, but many admitted it was a common practice. An article in the *Saturday Evening Post*[3] recounted the story of an operator in 1907 who listened in on many calls. She noticed a pattern of the types of calls throughout the day, from early morning conversations arranging for groceries to late evening calls of housewives checking in on when their husbands would arrive home from work. In 1907, it might have been one of the first recognitions of behavioral patterns based on phone call data. In a way the operator could predict the types of calls she would be asked to connect in each hour of the day.

The information switchboard operators could learn was serious business, and to demonstrate the phone companies' dedication to privacy, they did not allow the operators, who were overwhelmingly women, to marry husbands with information-sensitive jobs such as policemen, politicians, or government employees. Ironically, even though switchboard operators and phone companies learned the value and consequences of this policy nearly a hundred years ago, recently Amazon was discovered to be listening to Alexa voice recordings.[4] Amazon, which discontinued the practice, claims it was to better label voice commands and provide data for their natural language processing (NLP) algorithms. In other words, they were listening to you to help make things better for you, and maybe make it easier to order groceries.

Amazon is not alone in the data it can collect about you. All the large technology companies have a form of data collection that they meld together with other data they purchase to create an accurate data profile of you. If you have ever wondered how browsing for items online later results in a targeted ad for a similar item on your phone, perhaps in your Facebook news feed, then you know this is true. If you have a Facebook account, you can look in the Settings section to view and download all the data they have about you. It may be surprising to see every message, "like," uploaded photo, group and friend request all neatly assembled into a data file. Famously, some of this information was available more publicly as recently as 2016, when Cambridge Analytica, an online targeted marketing firm, used the information in political campaigns. The ads were targeted based on information they had obtained indirectly from Facebook without the users' permission.[5] The company came under investigation and ceased operations in 2018. Meanwhile, Facebook has been fined $5 billion and has worked to plug the gaps in its data infrastructure, adding more protective layers to its customers' data.

In addition to data companies performing self-regulation concerning data privacy, there are various governmental efforts too. In Australia there is the aptly named Privacy Act, in Europe, the EU has adopted the General Data Protection Regulation (GDPR), and in California, there is the California Consumer Privacy Act (CCPA). Each limits how companies can collect and use personal information and has led to some changes in transparency, such as Facebook and Google now making your data available for you to see. Earlier the notion of privacy was more focused on protecting your information from fraud and identity theft by protecting your credit and banking information. We were concerned with people running off with our money or impersonating us and ruining our good name when they wrote a string of bad checks. More recent privacy laws such as the GDPR are working on resolving how data are used in targeted marketing and a type of data deception that is not criminal.

Companies see the data of an individual as a valuable asset. Individual profile data can be monetized, traded, enhanced, and used for purposes *you* never intended. Companies like Facebook, Google, Instagram, and Twitter create fun online experiences, collecting data along the way that they later use to create a profitable digital image of their consumers. The GDPR introduced an interesting provision. It allows people to request their data be removed. The so-called "Right to be Forgotten" will remove all of your personal data from a provider such as Amazon. But before you do, consider there might be some benefits to you when those companies have data about you.

One of the things we are optimistic about is the future use of data applied in a responsible way to enhance our lives. Sure, this sounds like a too-good-to-be-true statement, but there are ways in which sharing data can be useful. Think about it this way. Hopefully, you have a doctor you see regularly for routine checkups. Over time your doctor takes notes about your visits, records any lab tests and perhaps most importantly gets to know you a little, noting your interest in healthy living and exercise and your love of Reese's cups. All those things together help create a type of medical profile for you, so when your doctor is making recommendations, they can consider all of that information as a whole. The fact that you share information with your doctor and they keep it in an organized manner helps provide you with better health care.

The same scenario can play out for your financial advisor or anyone else you consult with in your life on long-term important decisions. Sharing your data with Netflix, Google, Amazon, and others may have a similar type of benefit. Although Amazon is not going to solve your health care needs just yet, sharing information with them as you make purchases online, stream movies, and choose Kindle titles all can enhance your experience. Future online shopping is met with a more robust (and hopefully accurate) recommendation engine that suggests the right products, at the right time. Kindle books and Amazon movie choices as also more tailored to your liking when you share data, and even future movie productions can be based on consumer preferences. If you, and others with a similar digital footprint, prefer romantic comedies where the couple sails off into the sunset at the end, guess what? Amazon is going to use that data to write stories and produce more movies just like you want because, after all, that is what the data are telling them, and companies like Apple, Facebook, Netflix, Google, and Amazon did not get to be the companies they are without following the data.

Having all these data creates a personalization-privacy paradox. Companies need to demonstrate that the collection of private data provides benefits to their customers. Studies have shown if customers are aware of the data that is overtly gathered about them and know how it can benefit them, then they are more willing to accept it.[6] Many customers are unaware of how data are collected about them and being used. This type of *covert* data collection, especially location data powered by your phone apps, is a whole different type of data deception.

IT TAKES TWO (TO BE DATA DUPED)

Data and harvesting its value through data analytics is the new super-power in the analytics era. As a consumer, as a benefactor, and even just as a casual observer, your knowledge and understanding of how this works better prepares you to avoid being data duped. Data duping takes both a data duper—a person or other entity—and a data believer. Without these two, we would not need to worry about how data could be used deceptively.

Data can be used to influence and drive decisions. How will your decisions be influenced by information, analytics, and data? In this chapter, we have provided a foundation of data analytics and how it has evolved. As a more informed individual, we believe you run less of a risk of being data duped by having an understanding of the mechanisms to which nearly all of our economy is gravitating.

Unlike previous industrial revolutions, the analytics era is expanding rapidly. The first Industrial Revolution of the 1780s unfolded over a period of 70 years. Although it was disruptive in many ways, the speed at which it developed allowed individuals to adapt more easily. Manu-facturing jobs that relied on manual labor slowly disappeared, which gave time for people to learn skills they could use in the new factories being built. Time eased the transition. People gained an understanding of the new technology gradually. They built a certain amount of trust in it. We are certain some people distrusted the reliability and power of the steam engines propelling the factories in favor of the trusted-and-true methods of the past where they could depend upon manual laborers. Just like some people today do not trust Google Maps, but eventually, people came to see the benefits of the then-new era.

In contrast, the analytics era is currently having tremendous impact in just a few short years. New business models are emerging, like Insta-Cart for online grocery shopping plus an internet-connected refrigera-tor that helps you plan your shopping list. Complete job categories are being eliminated, and many routine and repeatable tasks are replaced either by technology or robotics. Examples include AI cameras read-ing license plates instead of tollbooth operators collecting cash and the robots Amazon employs to fetch items in their warehouses.

The media, like other business sectors, has been dramatically changed by data. Data we see in media—the news, Facebook, and marketing—are all subject to a new type of scrutiny. What data can we believe? How was it created and for what purpose? Is the data you are seeing random or deliberately presented to you? How are the various profiles of you created and used? Knowingly, we are relying on data more and more in our everyday lives. Data empowers us to make better decisions, but the deluge of information and the numerous places where we can encounter it also make the task of avoiding a data dupe even more challenging. At least now you know more about how the mechanisms of data work, so next we will explore strategies to help you prepare for the data revolution ahead.

KEY POINTS

- The term "data scientist" is relatively new and is often used interchangeably to describe a wide range of data-related jobs. Many of the techniques of data science have existed for a long time and were applied throughout history.
- The current analytics era is as disruptive and powerful as the original Industrial Revolution. Several factors have converged in the past few years creating a tipping point for the explosion of data and its application. Factors include exponential growth in computing power and a decline in costs, cloud computing, and immense amounts of data collected from nearly every smart device.
- The analytics era is prominent in emerging business models and the dominance of technology companies—five of the top twenty largest companies are technology firms.
- Unlike previous industrial revolutions, the analytics revolution includes direct participation by consumers. The data interaction between individuals and companies is yielding more sophisticated computer-based recommendation engines and personalized solutions and products. Personalization based on too much individual information creates peril for companies, and privacy remains a challenging topic for those balancing the privacy-personalization paradox.

Data Defense Strategies

THE VALUE OF INFORMATION

In 1981, the US government created a team at the Pentagon called the Active Measures Working Group, which was focused on the Soviet disinformation apparatus.[1] The growing cold war threat by Russia was being countered by the US deployment of weapons in Europe. One report characterized the deployment of Pershing II ground-launched cruise missiles as operating "within 6 minutes of striking Russia."[2] Russia, unable to counter these weapons with those of their own mostly due to economic reasons, set out on a disinformation campaign to discourage European countries from allowing the United States to use their countries as staging areas. In addition to a war of numbers, it was also a war of words, such as a now discredited story of US involvement in the assignation attempt on Pope John Paul II in Italy in May 1981. They also succeeded that same year in influencing a million-plus gathering in New York's Central Park to protest the nuclear arms race, which later impacted policy votes in the US Congress.

It was a game of deception by numbers. If the Active Measures Working Group got it wrong, the Russians would drain their economy to build even more weapons. The cost to the welfare of their citizens would be devastating and at the same time present a threat to people in the United States and its European allies. Russia was about to have tremendous unchecked firepower and was ever more on the edge of a hair-trigger reaction to information. There was no room for error. The Soviets, preferring to build false narratives rather than costly weapons, were winning a war of data and misinformation.

This example may seem extreme in the context of data duping, but it demonstrates how far information, when unchecked, can change people's perspectives and perhaps even change the decisions they make every day. From exaggerated claims ("Lose 20 pounds without dieting!") to misleading headlines ("Millions of Lives at Risk!"), we are all a little susceptible to the risk of being data duped. There are things we want to believe, prodded along by our own biases and things that are easy to believe. However, without a few data defense strategies, the unbelievable may be all we know.

CONSIDER THE SOURCE

Data deception has probably gone on for millennia. From the days of cavemen, we can imagine those ancient drawings on cave walls may not have shown exactly where Zog and Ugg were winning their best hunts of the woolly mammoth. They had a reason for their readers to have a certain view. Likewise, in modern times, if you ask a local fisherman for the best spot to catch fish, you are more likely to get directions to a location miles away from where the fish are biting.

The source of the information, the basis for one's claim or call to action, is very important, especially if that source remains hidden, or at least obscured by the summary article you might be reading or the advertising you see. Academic articles are better at providing transparency for their conclusions, with many making their source data freely available online on GitHub, Data World, and other repositories. Notes of their methodology, exclusions, and treatments are often provided also, for those eager and diligent enough to dig into the data. Digging into that data is useful, and many times you will find the summary headline you read in the news is just part of the study and perhaps overlooks even the author's concerns about absolute conclusions (many papers have conclusion notes of additional items to be researched). Yet with all the academic rigor there are criticisms.

Beginning around 2005 some academic papers were written ironically about academic papers, specifically whether or not the conclusions in them were true. John Ioannidisis of the University of Ioannina and Tufts University wrote, "There *is increasing concern that most current published research findings are false*" and "*it is more likely for a research claim to be false than true.*"[3] Other papers have since followed, revealing "academic inaccuracies," and several papers have been retracted over the years. Commonly, as Ioannidis and others conclude, there are several factors such as flawed experimental design, small ineffective sample sizes, the bias of observers and data collectors, and a low threshold for statistical significance. "Statistical significance" is the term scientists use to determine if their numerical results showing a relationship between two or more things are merely chance or a result of some cause and effect. Traditionally that threshold is 5 percent or 1 in 20, meaning that if the data crosses this threshold, then scientists can conclude the relationship between two things is not just luck, but a real relationship. Recent critics of this tradition argue evermore complex studies such as the evaluation of the human body in response to medicines and therapeutic treatments require a much smaller threshold to prove the efficacy of the relationship.

Another example of the need to check the source is found in weight loss advertising. Weight loss is a serious business generating $72 billion in the United States in 2018.[4] Assuming roughly 25 percent of adults are on a weight loss program, that is nearly $1,000 per person each year. Further spending on weight loss is growing two to four times faster than the population.[5] Not only do weight loss providers want to sway your decision to purchase their product, especially amid strong competition, they also want you to make that decision based on numbers! Yes, indeed, most weight loss advertisements quote numbers as to why you should choose them. In 1953, Zsa Zsa Gabor (yes, darling, that Zsa Zsa) appeared in print ads exclaiming how you could lose weight *and* eat all you want, simply by eating a little candy called Ayds, losing "up to ten pounds with your first box." The phrasing of "up to" might have been the first clue to the astute data duper, since it also means it could be less than that amount. The bias of the reader, however, was more focused on the benefits than the claims of certainty.

Although the brand faded away, the idea did not. By 2013, Sensa was growing in popularity as a weight loss program. Just like its 1950s Ayds counterpart, it was an additive that suppressed your appetite, and thus, without any dieting, you would lose weight. Fantastic claims of losing 30 pounds, some even losing 50 pounds made the media rounds. Also just like the older campaigns, they claimed to have "clinical studies" to back up their claims. None can be verified from the 1950s, and Sensa's claim to have a peer-reviewed study by the Endocrine Society was refuted. One thing that improved from the 1950s was government oversight on false claims and advertising. In 2014 the Federal Trade Commission (FTC) assessed Sensa with a $26 million fine and barred them from continuing business. Interestingly, the FTC gave them a chance to go forward provided they could create a true clinical study to back up their claims about their product using data. Instead, Sensa declined and declared insolvency. Perhaps they were done duping people with data that did not exist.

The source of data and the decisions they are imposing on you can always be subject to bias. The bias can be unintentional or a result of deliberate deception. Either way, as the target of the information, a good strategy to avoid being data duped is to be aware of the source of information and if possible, their intentions. Does a charitable organization want you to make a donation or simply inform you of their cause? Does a newspaper want you to subscribe (for a fee) or present balanced information (or both!)? Does the local city council want you to approve the next public works project and the upcoming bond referendum that will pay for it? Does your project team at work want to sway your decision about moving forward? All of these are examples of the "intentions of the source," and one of the most effective ways to address them is to find other sources of information—confirming or denying the direction of the decision they are leading you towards.

KNOWN REFERENCE POINTS

If you have ever seen the movie *National Treasure*, you know information and deception can sometimes go hand in hand. In the movie the "clues" were meant to dissuade the treasure hunters from where to find it, in order to change their decisions from looking in one place to looking in another far away from the actual treasure. Of course, our heroes see past this and recognize a larger pattern of clues (data) and, like in all good Hollywood movies, find the treasure and save the day. This would not be possible in the story if our hero was not only following historical clues but also an expert historian. He has more information about the data in front of him because he knows about other events in history. The context and reference points to other things help him with his decisions, to see past deceptions and make better choices. Starting with known reference data can be a good countermeasure to misinformation.

Reference knowledge is an important tool in your data-duping defense. Let's take an extreme example. Say someone tells you that they bought a new car and it is superfast. So fast they claim it can drive across the country in less than a day. Starting with what you may know: A trip across the country is roughly 3,000 miles, and a day is 24 hours. Then, dividing those two numbers means a car would have to drive 125 miles per hour . . . nonstop. Is that possible? Maybe. A NASCAR vehicle can top out around 200 miles per hour in the right conditions. Although it may average less than this, it still seems capable. As an aside, NASCAR is an acronym for National Association of Stock Car Racing, but despite the debate from your family members on race day, there is very little that is "stock" about them. It may look like your Camry sitting in the driveway, but don't even think about driving it at 200 mph, that just won't happen for 500 miles straight.

The Indy cars are faster and lighter than the typical 3,000-pound NASCAR and can hit speeds over 230 mph. Formula One cars, with lighter bodies and frames and larger engines, can be even faster. So, given a little reference knowledge and a starting point, you can make a few deductions from your friend's claim. It is possible to drive very, very fast across the country in a day. But if you look over your friend's shoulder and see a Toyota Prius, well you just might want to call him out on it. Seeing is believing and that, too, is another data point.

ESTIMATING—FERMI'S BACK-OF-THE ENVELOPE ESTIMATION

Whether you are bringing understanding to the Thanksgiving table or trying to win over your co-workers, your ability to understand numbers is important. In the previous section, we showed how having a reference point on one topic can help you understand another. In this section, we will take this a bit further by adding a few calculations.

Enrico Fermi was an Italian-born physicist who immigrated to the United States in 1938 following his acceptance of the Nobel Prize in Physics. He is widely known for his work on nuclear reactors and is credited with creating the first functional reactor, called Chicago Pile 1, which was among the first major milestones of the Manhattan Project.

The Manhattan Project would later go on to develop the first nuclear bomb. Chicago Pile 1 was named as such because it was partly assembled from a pile of 45,000 ultra-pure graphite bricks stacked upon each other. Fun fact—being the world's first nuclear reactor, it was built with a bit of secrecy under the seating stands of the University of Chicago's Stagg Football Field. If you are the type of person who is squeamish about living near a nuclear power plant, imagine sitting down at one of those U of C football games.[6]

In 1945, when Fermi was working on the Manhattan Project, he was witness to the Trinity test, which was the first detonation of a nuclear device. Fermi was curious about how to measure the strength of the explosion. At an observation point 10 miles from the explosion, Fermi dropped pieces of paper from his hand at head height and watched how they fell to the ground before, during, and after the explosion's shock waves past him. He quickly did some rough calculations of the displacement of the paper by the shock wave and came up with a fair calculation of the power of the explosion. Thus, was born the "Fermi Estimate."

Although the research shows Fermi was often doing these types of back-of-the-envelope calculations, this example was by far his most famous. Perhaps it was the lack of Google-like resources to look something up or simply the absence of quick calculating computers that motivated him. It demonstrated that with basic measurement tools, such as the falling paper, some estimates, and somewhat easy calculations one could arrive at a reasonable number for nearly anything. This principle is embraced by many job interviewers who often will ask an applicant to make estimates. Mostly this is aimed at how well someone can form a problem, break it down into estimable components, and then present a reasonable solution. Base knowledge and accuracy are not what is being tested, and a good "Fermi" estimate is one within a couple of orders of magnitude. Let's take a look at how it works.

One day your uncle Dilbert, who is proud of his alma mater, proclaims there was a time when his football team was so good they scored 1,000 total points in a single season. They called it the "Season of 1,000" and it was a miracle season. Maybe this is true, maybe he heard it somewhere. The years have been long and dates are obscure like the rest of his memory. There is no way to look it up. Are you and your uncle being data duped by some football lore?

You know a little about football, but you have not memorized all the teams' records. No matter, we can approach testing this statement with a version of a Fermi-type estimate. First, we know the football season plays in the fall and is roughly four months long. Further, we know football games are played once per week, meaning there were likely 16 games. Maybe a few more—if they were that good there must have been some playoff games at the end of the season, so we will round it to 20 games.

How many points could have been scored on average for each game? People are invariably poor at estimating likely, probabilistic outcomes such as the typical score of a football game. Our bias comes into play, such as recent football games we may have seen or read about, and we do not have a full perspective of all the games played to calculate the average scores. However, what we are pretty good at doing is estimating ranges. If this was a winning team, what is the range of possible scores for any given game? We will make a range of a low of 3 and a high estimate of 120. It seems unlikely to us that a football team could score more than 120 points. Rather than taking the *average* of those two numbers, we will calculate the *geometric* mean. The geometric mean sounds exotic but is it quite simple to do using your smartphone calculator. Technically it is the square root of multiplying the low and high estimates.

Geometric Mean for two numbers = $\sqrt{(\text{low estimate} \times \text{high estimate})}$

Why use a geometric mean rather than an average? In real life, most ranges of data behave more like a right-skewed distribution—think about people's heights, household income, length of passwords, distance traveled from your home in a car, number of rainy days in a row, and so on. Right skewed means there are more values closer to your low-end estimate than the high end, because, after all, your high estimate is the most (reasonable) extreme value you could come up with.

Calculating the geometric mean using our estimates gives us:

$97 = \sqrt{(3 \times 120)}$ *or about 20 points per game*

Again, we are rounding to make the math simple. Multiplying this by the estimated 20 games gives us a cumulative total of (20 points × 20 games) = 400 total points. It is not looking good for the claim of the "Season of 1,000 Points."

Now suppose you are working through this estimate with your uncle Dilbert. He says "wait a minute" and now recalls hearing they were such a good football team they never scored less than 20 points in a game. What would this do to your estimate and does it make the story more believable?

$$48.98 = \sqrt{(20 \times 120)} \text{ or about 50 points per game}$$
50 points per game × 20 games = 1000 points!

Well, now we have something to cheer about. See how a little bit of additional information can improve your estimates and test dubious data? It seems quite possible the "Season of 1,000 Points" could be true. Go Tigers!

COMPOUNDING RATES

If you are familiar with the "time value of money" concept, you might know about how compounding rates work. The time value of money is a type of calculation to show how money increases or decreases as it is moved through time. It is the change in value when either someone is paying you interest or you are paying someone else. In a simple example, if you put $100 in a bank deposit account and the bank agreed to pay you 5 percent, at the end of the first year you would have $105, calculated as $100 × 1.05. Keep it there a second year and you would have $110.25 ($105 × 1.05). A simple formula makes calculations easy for any given number of years:

Future Value = Initial Deposit × (1+ interest rate)^{years}

Compounding rates also work in reverse, allowing you to go backward in time. For example, if someone asked, "What is the present value of $110.25 two years in the future discounted at a rate of 5 percent?" we could apply the formula this way:

$$Present\ Value = \frac{Future\ Value}{(1+Interest\ Rate)^{Years}}$$

$$\$100 = \frac{\$110.25}{(1+0.05)^2}$$

Not only is this useful for calculating money, but it can also be applied to other things that increase or decrease over time. Being able to use these calculations can help you with simple estimates. When someone mentions electric vehicles (EVs) are becoming more popular in the United States and soon we will all have one, how can you test this claim? If the current number of EVs produced is 330,000 and the production has been growing 30 percent each year, how many will we have in 10 years?

$$\#\ of\ Electric\ Cars\ in\ 10\ Years = 330{,}000 \times (1+0.30)^{10\ years}$$
$$= 4.5\ Million\ Cars$$

With 275 million adults in the United States, it is unlikely we will *all* have one in the next 10 years even with this incredible growth rate. How long, then, will it take? Assuming our population stayed the same and all-electric cars remained in service, we can use this same technique to estimate it will take about 25 years to reach 275 million electric cars.

Here is another trick that uses compounded growth rates. Each year about 11 million vehicles are made in the United States and the production growth rate is roughly 1 percent.[7] As we showed above, we can calculate the number of cars produced in the past by discounting the current year's production number by 1 percent. And if we wanted to know how many were produced 10 years ago, we could apply the formula. Further, if we wanted to estimate the *total sum* of cars made in the last 10 years, we could perform the calculation 10 times, once for each previous year, and then add them all up. Simple! Now, what if we wanted to know the total number of cars produced *ever*, since the beginning of time?[8] When the growth rate of an item is even through time, we can estimate the grand total as:

$$current\ production \times \left(1 + \frac{1}{rate}\right)$$

With the current production of 11 million cars and a slow growth rate of 1 percent annually, this works out to a total of 1.1 billion. In other words, since the beginning of time, the United States has produced a total of 1.1 billion cars.

HALF RIGHT

Years ago, we knew a manager who was good at making in-the-moment estimates. This manager was also good at recognizing skepticism when it came to others' acceptance of business analytics, which at the time was still a little-known discipline. The manager could sense the leaders we were trying to influence were a bit leery of how a bunch of math was going to help them make better decisions. After all, what did the math tell them that their years of experience, working face-to-face with customers, could not? To avoid distilling the mathematical models—a process of several follow-up meetings and perhaps ultimately delaying the decision, the manager's approach was simple: What if all this math *was nearly right*, and some of what it was showing *was* believable, if not convincingly true? What if it was half right?

A conversation that began as a tour de force of numbers and probability theory, and academic terms such as "Gaussian distributions" and "Bayesian contingent probabilities" and "the value of geometric means," was quickly dispensed. The point was to put the math aside and focus on the decision. With the decision front and center, we could move on to how much math, or simply stated *numerical evidence*, was needed to make that decision. What if it was half right? Often what we learned is we did not need to have perfect information or a perfect model to interject some analytic insight into decisions.

This lesson in simplicity can be an added tool to your data defense toolkit. When confronted with data related to a decision, how much information do you need? And how much do you need to know if the data are being used in a deceptive manner?

You may be confronted one day by a decision where you do not have all the information. Or someone may be challenging you about a decision that cannot be made because of missing information. The spirit of a Fermi-like estimate is to get a quick approximation, to be able to do some mental math and arrive at a conclusion, or to make a decision.

A fair way to approach this is to ask after you have an estimate if the amount was half right, could you still agree? Would you need to further refine your numbers? If someone is trying to deceive you with data, how much more would you need to know? Is their numerical argument half believable? If so, then there may be no need to dive deeper into the numbers, to challenge or further explore. If not, you are at risk of being data duped.

Of course, there are times when accuracy matters more than others, when you need to go further. The purpose of a "half right" approach is to quickly dispel the deceptive and spend time on the more constructive. When there is an 80 percent chance of rain, when a survey says people agree 75 percent +/− 4 percent, when a drive time is 4 hours, then the decisions are easy. If these numbers were smaller with more uncertainty, you might reconsider your decision.

TOO GOOD TO BE TRUE

"This new car will run for 250,000 miles without needing a major repair."

It is unlikely any major automotive manufacturer would make such a claim, but maybe an unscrupulous salesperson might. Or maybe your uncle Dilbert, who is proud of his recent purchase, just heard it that way. Is it too good to be true?

Typically, too-good-to-be-trues (TG2BTs) are obvious. And claims such as this are easy to spot. We know to steer clear of making dubious data decisions. But what about when it is a little more subtle?

In 2009 Kellogg's Rice Krispies packaging claimed that since their cereal provided 25 percent of your daily recommended nutrients and vitamins, it helps support your immunity. The word "immunity" was in giant letters on the box. Moreover, it was during the height of the swine flu scare. It sounds reasonable but maybe just too good to be true. Eat a bowl of cereal and boost your immunity. Easy, right? The Federal Trade Commission did not agree and in 2010 required the marketing campaign and packaging to change.

How do you know when something is too good to be true and how to verify it? There are a few approaches and usually, we look at economics, motive, and comparisons.

Economics is always a good place to start for TG2BTs. Depending on their claim, you should be asking how can they make any money? When an automotive dealer is offering a $5,000 rebate on a new car, how can they afford to give away that much money? Usually, it requires you to sign up for their financing and there may be other transaction fees to make up for it. Or simply, the vehicle is overpriced in the first place.

Google and Facebook made famous the phrase "If it is free then *you* are the product." This was largely a result of the huge revenues each of the companies earns as a result of profiling the data you freely share with them. Facebook's 2019 advertising revenues were $70.7 billion and have increased nearly 300 percent just since 2015 when revenues were $17.9 billion. Current earnings divided by the 1.6 billion active users means the value of each person sharing their data is $43. The economics are certainly in their favor, and some could argue Facebook should be paying its customers for contributing to their windfall revenues.

Another example is Amazon. Of course, their value is as a retailer and distributor of real products, but they are also very much in the data business and recognize the value of their recommendation engines to the other retailers. A recommendation engine is a data tool that uses information about you, such as your past purchase history, to predict other products you would buy. These data can help retailers know which products to offer when and at what price points people are willing to pay. Recommendation engines are so valuable, in fact, that in 2020 Amazon is expected to earn more than $13 billion in advertising revenue, which is an increase of 23 percent over the previous year.

In addition to your purchase history, Amazon is also collecting data you freely share through your Alexa smart speaker device. In 2017 Amazon was selling Alexa for $180 and, according to a report by HIS Benchmarking, it cost $57 for parts and manufacturing, which means at the time Amazon was charging a *premium* of $123. Prices have since dropped to $75 for Alexa and just $29 for their smaller Dot model. However, there is some rationale that says the product should be free if you are agreeing to share your data, and perhaps like Facebook and Google, they should pay people for using it. Maybe at the moment, this is just TG2BT for them, rather than you.

What is the motive of a too-good fact or piece of data? Similar to "Consider the Source," one needs to think about possible motives for something that seems too good. In data, something that may be too good to be true are streaks, or improbable streaks. Streaks such as tossing a fair coin and getting heads many more times than tails or a winning streak by a particular athlete or sports team. Or a stock market picker with an unusual ability to pick winning investments, who then offers you a chance to put some of your money into the market.

Perhaps the best TG2BT story is the one famously named after Charles Ponzi in the 1920s. His investment company started on a simple idea: different rates for international postage between countries, specifically, the International Reply Coupon (IRC). An IRC could be exchanged for first-class postage in several countries, and it was common courtesy when writing for someone's reply to include an IRC so they were not burdened with purchasing postage. Ponzi's idea was to purchase IRCs in European countries at a lower cost than the US postage stamps they could be exchanged for and then sell the US stamps, netting a profit. The margins were thin and the logistics too impractical, especially in 1920. Ponzi needed cash to get the plan off the ground and promised his new investors extraordinary returns of 50 percent on their investments in just 90 days. Later he promised the same returns in only 45 days. Perhaps overrun by the enthusiasm of his investors, who were eagerly handing him cash, and his own greed, he soon abandoned the stamp idea. He discovered by paying some early investors "returns" using the cash from his new investors, the demand rapidly increased. In a matter of a few months between late 1919 and August 1920, he had established several offices for his Security Exchange company and pulled in more than $20 million. Adjusted for an average inflation rate of 2.59 percent, it is a staggering $258 million in year-2022 dollars. He continued to funnel money from the new investors to payout the "returns" of the established investors. The model was not sustainable. Eventually, prompted by a newspaper story, investors demanded their money back. The money was gone and so soon was Ponzi—off to federal prison. The scheme left a mark in history and coined the term "Ponzi scheme" for this most famous fraudster.

Sadly, that was not the end of Ponzi schemes. Between 1991 and 2008, Bernie Madoff gathered $57 billion in investors' money. Similar to Ponzi's scheme 80 years before, he used money from recent investors to payout others. Madoff testified that he was not even attempting to invest the money and simply kept it all in a bank account, adding and withdrawing funds as needed. The net loss to all of the investors is not exactly known but is estimated to be between $12 billion and $20 billion.[9] It has been noted that without the recession beginning in 2008, which increased people's need to withdraw their investments, Madoff's scheme might have gone on a lot longer.

The data were there and, in fact, in May of 2000, Harry Markopolos, who was a portfolio manager for an options trading company, provided an eight-page complaint to the Securities and Exchange Commission (SEC) after he had done some analysis of Madoff's numbers. He concluded that the consistent returns of Madoff just were not possible. Markopolos was not deceived by the numbers like so many others. Later in 2005 he resubmitted his list of concerns and described the performance of Madoff's hedge fund:

> I presented 174 months (14½ years) of Fairfield Sentry's return numbers dating back to December 1990. Only 7 months or 4 percent of the months saw negative returns. Classify this as "definitely too good to be true!" No major league baseball hitter bats .960, no NFL team has ever gone 96 wins and only 4 losses over a 100-game span, and you can bet everything you own that no money manager is up 96 percent of the months either. It is inconceivable that BM's [Bernie Madoff's] largest monthly loss could only be –0.55 percent and that his longest losing streaks could consist of 1 slightly down month every couple of years. Nobody on earth is that good of a money manager.[10]

There it was. "Definitely too good to be true." It would take nearly three more years before, in 2008, the SEC shut down Madoff Investment Securities, LLC.

As the old rule says: If it sounds too good to be true, then it probably is.

PATTERNS AND BENFORD'S LAW

We are data people and we have some peculiar habits. Plop a set of numbers down in front of us and we will start to look it over in ways that others might not. No matter if it is the weekly temperatures in

Reykjavík, Iceland, or the balances in a savings account. We will look at the averages and medians, the biggest number, then search for the smallest. We will sort the numbers in some logical manner—perhaps biggest to smallest or by date. Even when not asked we will look through the numbers searching for patterns, because patterns reveal information a bit like a farmer looks to the evening sky to get a read on tomorrow's weather.

Inevitably there will be something that makes us ask "why?" When inspecting our numbers, we might wonder: Are the numbers getting bigger or smaller over time? Are there any numbers that stand out? Are there recurring numbers? For example, why did the temperatures in Reykjavík increase? The answer might be as simple as, it was spring-time. Whenever we see numbers that are pointing to a trend or seem out of place, we should ask if there is a reason, like the seasons in Iceland or the large purchase you might have made that decreased your savings account.

Sometimes patterns in numbers are easy to spot and later we will talk about data visualizations as one tool, but another lies within a deeper look at the numbers called Benford's Law.

It is incredibly difficult for someone to just make up a set of numbers, like the Madoff customer's falsified investment statements. Humans are not very good at making up numbers that fit into a natural and believable pattern. Too often we are guided by our internal bias and reference points. Take this example. Pick a number between 1 and 50. Any number up to 50. Wait, make sure the number has two digits up to 50 and those two digits are odd numbers and not the same number. Two different odd digits of a number under 50. Now lock that number in your memory. Is your number 37? You may have seen this mental trick performed before and know about 70 to 80 percent of people choose 37 or a similarly close number such as 31, 35, or 39. What this dem-onstrates is when tasked with making up a list of random numbers, we humans are not so good are randomizing. Neither was Bernie Madoff.

Naturally occurring numbers in a wide range of data sets from bank-ing transactions to the length of the world's rivers have been observed to follow a pattern. The pattern known as Benford's Law shows the leading digit is most likely to be 1 (about 30 percent of the time) with the second most common leading digit as 2 (about 17 percent of the time) and so on in diminishing percentages.

The pattern was first discovered by Simon Newcomb in 1881 although Frank Benford, a General Electric electrical engineer and physicist who wrote about it in 1935, is more often associated with the principle, also as known as the first-digit law.

Bernie Madoff's investment company was required to publish monthly returns and as Markopolos pointed out they were amazingly consistent, having only a handful of negative returns. After all, Madoff was making up his investment trades after the fact, so how could he not avoid being right most of the time? How did his pattern of monthly returns compare to the expected Benford's Law? Following Benford's Law, about 30 percent of the returns are expected to begin with the number "1" (note: 11 percent and 0.1 percent both count as "1" since leading zeros are not considered). However, the Madoff returns favored "1" as the first digit in 40 percent of returns and used "2" through "5" less often than would be naturally expected.

Benford's Law is not a law of science—the numbers do not have to follow this pattern. It is possible the Madoff anomaly is a result of coincidence and the result is real, but this is a good data defense tool that is easy to use. The point is, if investors and regulators had observed this pattern, they may have started other investigations and they may have discovered the scheme sooner than they did.

CONVENIENT CORRELATIONS

A correlation is when two things move together. Like dancers gliding across a dance floor, there is a relationship between them. Step for step they move and twirl and although the steps of each dancer are not exactly the same, they are linked and generally moving in the same direction. Their placement on the floor if it were plotted with x and y coordinates would tell us there is something special about their movements. They are different yet we would know that one seems to be influencing the other or vice versa. A move to the left also draws the other in the same direction. A move to the right, the same. As people in tune with numbers like we are, we could imagine the measurements of each step—their appointment in time and space as they touched the floor and sprung off into another direction. And with a little effort, each of those measurements might tell us a lot about where the next foot will fall. The relationship between dancers, space, and time could let us know how well they are working together. And if they are not working together, not dancing in step, we could know that too.

In numerical terms, a correlation is the degree in which a pair of data points is related. This might be the relationship between the height of children and their age, the temperature with the time of day, or the placement of one dancer's feet next to their partner's. Technically it is measured by calculating the corresponding variance of the pair of numbers. This roughly measures how one set of numbers varies or changes its average compared to the second set of numbers. You do not need to know how to do the math since there are many tools, like Excel, that can do it for you.

Let's take a simple example of children's height by age.[11] If we plotted the data, we would easily see a relationship, a correlation, between age and height. As children get older they often get taller. If someone made the statement that the two go together and showed you the data, you would absolutely agree. If they also provided the strength of the relationship known as the correlation coefficient, you would need to know a little more about its interpretation. Correlation ranges from negative one to positive one, meaning when a pair of variables are moving in an opposite, but related direction, then their correlation will be closer to negative one. When moving in a similar positive direction then the measure will be near positive one. In this example the correlation coefficient measure is 0.997, indicating a very strong positive relationship—as children get older, they certainly get bigger, on average.

Correlation is a linear measurement, but the data do not always have to be in a perfectly straight line. In fact, as the pairs of data deviate from a straight line so, too, will the strength of the measured relationship. A weak correlation value will be closer to zero. Thus, when one states there is a strong relationship between x and y, the measurement should be closer to positive one or negative one.

Hopefully at this point correlation seems intuitive to you and is seen both as a method to prove your point with data and a basis for how to build data defense. When someone else is trying to prove their point, you will be more aware of what is being measured and how sometimes it may be misleading. But how can that happen?

In 1973, English statistician Francis Anscombe created and published "Anscombe's Quartet."[12] His intent was to prove a point—in addition to measuring a data set's basic descriptive statistics, one should also plot them in a graph. Otherwise, the basic measures themselves may be deceiving. Apparently, even in 1973, Anscombe was concerned about being data duped.

To demonstrate, Anscombe contrived a most interesting set of data, as shown in figure 3.1. Actually, four sets of data each containing X and Y data points. In each set, the average for each set of X data points was the same and likewise for the Y values. Other statistical measures of the four data sets including correlation were *exactly* the same. Based on the classic statistics alone each data set seemed identical. But they were not. Given what we just learned about correlation, we would also conclude within each data set there was a very strong relationship (r=0.816) between X and Y. In fact, for at least two data sets this was true. However, the graphs of the data points show the influence of the outliers. **Beware of the influence of outlier data points.** In data set IV, we can see by removing just one outlying data point there would be zero correlation! How quickly our conclusion about the relationship between X and Y could change. If this can be done with a small data set, imagine a much larger one. Imagine only being provided the correlation and not

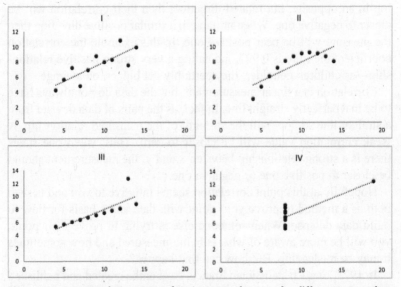

Figure 3.1. Anscombe's quartet demonstrates data can be different even when all the traditional statistic measures are the same.

the graph or the data to be inspected yourself. It might be easy to be data duped. Which brings us to "convenient" correlations, those that are so good and align so well to support a position. By now you know enough to be a data skeptic.

Headlines are among our favorite sources for deceptive uses of "convenient" correlation. In 2020 a *Sport Illustrated* article titled "Inside the Correlation between College-Town Infection Rates and Football Fan Attendance" referred to "uncontrollable COVID levels." This caught our attention because it demonstrated another important principle about correlations. Even when there is a strong numerical relationship between a pair of data points, the change in one does not always *cause* the change in the other. The relationship could simply be coincidence, or often that they are both tied to a common third influencing factor. Thunder and lightning do not cause rain, but often they are observed together. They have a high observed correlation because they are both a result of clouds. The adage among data scientists is "correlation is not causation," however, let's take a look at this further.

Unfortunately, the *Sports Illustrated* article did not provide the data behind the research, and other than a scatterplot, it did not provide any further statistics. The implication is there is a high correlation between the number of COVID cases and the number of seats at college football stadiums and some sort of associated recklessness this promotes among fans. The accompanying scatterplot did show what appeared to be a positive correlation between the current number of reported COVID cases and the total attendance at the school's respective stadiums. However, there are a few data deceptions.

First, the correlation is not explicitly measured. Also, the size of the schools varies and larger schools with larger football stadiums also tend to have a larger number of students and a larger number of COVID cases (whew, that was a lot of "*larger*" but we think you get our point). It appears the relationship between COVID cases and seats in football stadiums is more likely a function of simply having more students and not a result of their love for football. Next is the fact that the football season had not begun at these schools at the time the article was published. Also, many of these schools either would not be hosting fans in the stadium or will be limiting the number who can attend. If they are trying to show cause-and-effect from attending football games, then this study would be more meaningful after a few weeks of gameplay.

Does football cause COVID? We do not think so. We do agree being in close contact with people can contribute to the spread of the virus, but that is another story and not the one *Sports Illustrated* was trying to conveniently demonstrate with their article.

Why is this useful? Correlation can be a useful tool to show a relationship between two things. It can show a cause-and-effect relationship, too, when one exists. And to the data duper, it could also be misused to show a relationship when none really exists. Just like a person randomly walking across the dance floor while others dance together in step, his walk may look like he is participating when he really is not. In other words, the coincidence of a pair of data points moving in similar ways can look a lot like data evidence when it is nothing more than random luck.

DATA DECEIVE, NOT JUST PEOPLE

Data do not show any emotion. They are neither bad nor good, although for many in the analytics profession we hear a lot about "bad data" and "data for good." Data for good are more of an archetype for how many well-intentioned people are applying analytics for a greater moral purpose like reducing disease outcomes, solving hunger, and improving the planet. Bad data may just be poorly curated, messy, and disorganized, but trust us, the data are not mean, evil, or mischievous on its own.

When you encounter bad data, it is most likely the result of an error, a mistyped value, or a problem in the way the data were translated and stored. The first rule of defeating bad data is to examine them for values that are outside of the expected range. For example, if you are measuring outside temperatures in Celsius, do you expect any values to be beyond a range of -50 to $+50$? Certainly not beyond -100 and $+100$ unless you are measuring the temperature on Mars.[13] When you do find errors in your data, you can toss them aside and you will be well on your way to cleaning up the data.

Another type of "bad data" is one that just sits there and can deceive all on its own. Data can be minding its own business buried deep in your data set, or drifting through a cell in your spreadsheet. Just like tripping over a rock on a hiking trail, the rock did not intend you any harm, it was at the intersection of you and it. Data can be like that too. It can be out of place with the rest of your data, and we often refer to these "odd rocks" as outliers.

In broad terms, an outlier is an observation, a data point, that is well beyond the values of the others. There is no specific law to define an outlier and it is a bit of where science meets art, or at least artful interpretation. Drawing a picture of your data, perhaps as a scatterplot, may make identifying outliers easier to visualize as they would be the dots that extend beyond the cluster of others. If you have the opportunity to examine the data yourself, you can look for the largest and smallest values. There are many tools to find the extreme values and some are as easy as dropping a data set into Excel.

Why is it important to know if there are outliers? When you are faced with a decision, interpreting data, or trying to assess the information you have been given by others, you will want to ask about extreme observations of the data, aka outliers. Their impact may be great as they weigh on the averages or cause you to believe a relationship exists when it does not. Like Anscombe's quartet demonstrates, a few outliers can dramatically change the outcomes of our data-dependent decisions.

In addition to bad data and a few outliers, another way data may deceive on its own is due to a skewed distribution of the data. When a distribution of data, such as class grades, is graphed in a histogram based on frequency, it may resemble the "bell-shaped" curve you might remember from grade school. This type of distribution is a normal distribution, which means most of the students' grades were near the average and an equal number were below and above average. A skewed distribution when graphed may look more like a ski slope. When it is skewed to the right, called a positive skew, it means a small number of students did exceptionally well, while most students had lower grades. A skew could also be to the left, a negative skew, meaning most students did well but a few did poorly. Similar to outliers this is important when describing the average value and making conclusions. With data skewed one way or the other, the average (also known as the mean) will be "tugged" in the direction of the skew. Take an example of a particularly hard grade school exam, where most students scored 65 percent. If a few smart ones have perfect scores of 100 percent, then the average would be more than 65 percent, and if the difficulty of the test was judged only on an average score it may not look too tough. Remove a few of the top scores, and the average suddenly drops, resulting in a different conclusion about the difficulty of the test. Averaging data into the mean is useful although it often does not tell the full data story, which is why there is another important measure, named median.

MEAN VERSUS MEDIAN AND WHAT YOU SHOULD KNOW

In our experience, when someone is asking about an average or mean, they are looking for information about the "typical" item that is measured, and they often are about to make a decision on these typical groups. Median is the "middle of the road" measure of your data set and may be more representative of your typical observation, especially when the data set is skewed to the left or the right. Specifically, the median is the middle value when your data are sorted from smallest to largest.

Let's look at how much money people have saved for retirement. Your uncle Dilbert may say he has read a few headlines recently about the average amount people have in their 401(k) retirement accounts and things are looking pretty good. He references the recent study on how America saves[14] and shares that the average balance is $102,000. Sounds like a lot of money, but as you are becoming a data skeptic especially when broad claims are made, it should make you wonder if $102,000 is "typical." Do a lot of people have balances similar to this amount? Is average a fair assessment? If you are a product manager or a policy maker, knowing where most people are in their preparation for retirement would be very helpful. However, if you rely solely on the average or the advice of your uncle Dilbert, you may be in trouble.

Looking a little further we note the median value of 401(k) accounts is just $29,000. Whoa. That is 3.5 times smaller than the average! Again, the median is the middle value, so this means half of the accounts are *below* this amount and half are above. With 5 million accounts in the study, that means there are 2.5 million accounts with balances under $29,000. All of a sudden $102,000 does not seem very average or typical at all, and the difference between the mean and median implies the data are skewed with a small number of people having much larger balances. But if they do, then what causes the data to be right skewed? The answer is age.

Age has a big influence on the balance of retirement accounts. When you are younger and your income is lower, it is more difficult for you to put money into savings. Hopefully, most people start saving for retirement early in their working lives and if they do around the same time, then it makes sense that older people who have been saving longer will have accumulated more money. A 60-year-old who has been saving since they were 25 will certainly have more than a 40-year-old who started saving at the same time, not to mention the extra 20 years of compounded growth. According to data from the Federal Reserve, the median balance of the 65+ group is more than seven times larger than the 25-to-34-year-old group.

Having considered how the data by age influence balances, we would also have to consider it is likely there is also a skew within similar age groups. Not all 55-year-old people save alike. Household income and perhaps access to retirement accounts (not to mention the discipline to sacrifice a dollar today for one tomorrow) will also impact the balances and create variation among households. In the end, the original simple statement about "average" retirement savings becomes complex when we think about how retirement savings are not equally spread across people and how some may have considerably more than others. If we had the opportunity to dig further into the data, we may find more factors that impact the balances.

Knowing a few basic statistics about the data set can help you assess how much faith to put into making statements about the average. We may discover among retirement accounts there are very few that are exactly $102,000. Given the difference in mean and median and the skew of the balances, we would expect there are many more that are closer to the median value of $29,000 than the average. This would be a game changer if you were about to make a decision about the "typical" account holder.

DATA MODELS AND REGRESSION

Terms like data models and regression can both be intimidating and confusing, especially if you do not regularly work with data. Regression is a method of using data to create a model that might help explain something like the relationship between the duration of a storm, temperature, and the accumulation of water on a road. When you hear someone using a data model to convince you of something, you might ask how it could result in a data dupe when they are using such world-class analytic techniques. The answer is it's just like how a carpenter can make an uneven table despite having all the best tools. It is not about the tools. Tools, especially good ones, can paint the illusion the work must be great. However, it is about how well they are used that really matters. And that guide applies to data models too.

When someone presents a data model, they are presenting evidence to support their position. It is our job throughout this book to help you evaluate that evidence. Ask appropriate questions and know just enough about the topic not to be data duped.

What is a data model and where might you encounter one? The latter part of the question is easily answered with "everywhere." Analytical data models are ubiquitous in our lives. Predicting how long it will take us to drive to the grocery store, autocompleting your text messages, and even suggesting which restaurants you might like are all examples. Let's begin with a simple example.

Perhaps you are in the market for a new car, or at least a new-to-you-car, which is to say "pre-owned," and you want to know what is a fair price. The price of any used car has something to do with how old it is, how many miles it has been driven, its general condition, and perhaps some other factors. Intuitively this makes sense, but the data, and a model that uses that data, can help you be more precise. How much more should you pay for a car with 20,000 fewer miles or one that is a couple of years newer? A model, based on the data, assigns weights to factors so that the model closely matches the data. In other words, it will help you determine how much the price *depends* on the car's mileage, age, condition, and the like. And in fact, in statistical language those factors are called the *independent* variables, and the price is called the *dependent* variable.

If you are new to the concept of data models, you may view this as a bit of magical math. In reality, it is not complicated. The idea of data models dates back to the early 1800s when both Adrien-Marie Legendre in 1805 and Johann Carl Friedrich Gauss in 1809 independently developed the concept. You may recall Gauss is also the namesake for that bell-shaped "normal" distribution you might know from your grade school tests, when grades were adjusted to "the curve."

Back then mathematicians may have seemed like magicians to us in the modern day because they calculated data models and their formulas manually using paper and pencil. Again, the technique is not difficult but overly time-consuming, especially when the model involves many data points. Even by the 1950s, this task was labor intensive. Although the mechanical calculator had become a mainstay product, it still took people to compose and input the numbers and to record the results. When needed on large-scale math problems, the data were divided among many people in a room, who were referred to as "computers." Unlike our point of reference today, computers were people doing calculations on electro-mechanical calculators like the Friden STW, all sitting at adjacent desks where results could be recorded on paper and handed to the next individual for the next set of calculations. The room must have been noisy with the clacking and hum of the machines—one could literally hear math being done, which depending on your perspective was the sound of a rickety train on the tracks or a concert of wonderful solutions. Either way, it is very different from the computers of our modern era. The computers of the 1950s led to mainframes, which by the 1970s took about 24 hours to calculate a data model. Of course, today using your simple laptop you can get results in mere seconds. But first, we need to understand a few things about the formula for a specific type of data model called linear regression. It is named this because it uses data to create a line and the line explains the relationships between the inputs, such as car mileage, age, and so on, and the output—the price of a car. The simple formula for a line is:

$$Y = a + bx$$

where "a" is the starting point that increases at rate of "b" multiplied by some factor "x," resulting in "Y." Our instinct leads us to believe the price of a used car increases with age as a factor (newer cars are more expensive) and decreases with the mileage as another factor. There is some fixed amount, call it the starting price, that is adjusted for age and mileage. Adopting the line formula from above, these two factors could be written as:

Car Price = New Car Price + weighted factor × age + weighted factor × miles

We will begin with the age of the car using a sample data set of used cars offered on Craigslist from across the country (see figure 3.2). Naturally, the price of a car decreases with age, thus the downward sloping line. The math that is going on here is to fit a line that goes through, or closest, to most points. Technically this is measured by the squared distance from a particular data point to the line. The line where the sum of all those squared distances is the smallest is the best line and defines our equation. The figure shows when age is the only factor the price starts at $29,934 and you then subtract $1,432.50 for each year in age. A 10-year-old car then would be estimated as $29,934—(10 × $1,432.50) or $15,609. We can see from the scatterplot our predictive line does not go through all the points and further, there is a range of prices, even when the ages of the cars are similar. This tells us there is more to pricing a car than just its age, so do not be deceived by anyone who tells you otherwise. Let's look further.

Next, we will look at the influence mileage has on the price of used cars. Similar to age our scatterplot shows a decrease in price as the miles increase. When miles are the only variable, the expected price is $27,718 and decreases by 12 cents for every mile driven. A car with 100,000 miles would be $27,718—(0.1204 × 100,000 miles) or about $15,678. Like age, we still see variation among cars with similar miles and this, too, tells us miles are not the only thing driving the price. In the spirit of being simple, let's keep it to two variables and see what it looks like when these are combined to create a model with these two data points.

Now we know for every mile driven we expect the value to decrease by 12 cents and for every year in age a decrease of about $1,000. But what do we get if we put both age and mileage into our model simultaneously? That model follows:

Car Price = $32,250 + (−$1000 × *age*) + (−$0.07 × *miles*)

The data tells us a used car with zero miles and zero age (i.e., new) is worth $32,250. From there, subtracting $1,000 for each year of age and seven cents for each mile driven yields a price similar to what you would expect to see listed online. And just like that we have a predictive model. Predictive in the sense that if someone asked us what list price should we ask when offering a used car for sale and only provided the car's age and miles, we could give a reasonable, data-supported number. In fact this is exactly what occurs when you post an item for sale on eBay or some other site when they recommend a sales price. Of course, their models are a little more robust, but the same general concept applies, and it might be very useful when you are selling things you find in your attic and you have no idea what they are worth.

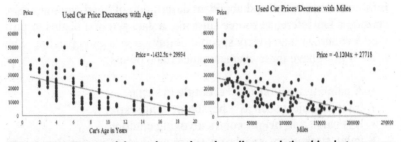

Figure 3.2. Data models can be used to show linear relationships between car prices, age, and mileage.

WHEN THE FIT IS TOO PERFECT

Now that you know a little more about how a data model is built and how it is useful, how then can it also be a tool of data deception? The answer is in a concept called "over-fitting." As we saw above, a model attempts to create a formula that "fits" most of the data. In our simple example we could easily see with a scatterplot that it did not intersect every data point. It was not perfect. However, by using some higher-level math and allowing our modeled line to bend and turn to connect with more points, we would have resulted in a much more complex formula. One, in fact, that *only* fits our unique data set and would be rather useless when applied to tomorrow's listings of cars for sale. An overfit model is one that overdescribes the data. Sometimes too much detail and too much description even when done with numbers is not helpful.

How could this be possible? How could too much information describing a set of data be a bad thing? Consider this analogy of someone describing a painting. Not a particular painting but more broadly the definition of a good painting.

Two friends, Yvonne and Emilie, are talking about art. Yvonne is interested in studying art and for whatever reason has never seen a good painting. She wants to know the definition of a good painting and how best to identify one. Emilie is a curator of a museum exclusively dedicated to Vincent van Gogh paintings and, having worked there so long, she has forgotten much of what any other paintings look like. Van Gogh paintings are particularly distinctive in style and unique in the painting world, as he was the among the most influential Postimpressionists in the 1880s. His style and subjects are unmistakable, which leads to

Emilie's rather biased and abundant description of good paintings. She considers her reference source (just like a data set) and begins to provide Yvonne with an overly specific definition of a good painting. She notes it must have these characteristics (data points):

- A painting is **square or rectangle** in shape
- It has a subject such as a **person or landscape**
- It is painted in **bright colors** invoking emotion
- Paint strokes are **swirls or short strokes** of contrasting mixed colors
- Paint is rigid with a **texture** called impasto
- Subject is often **distorted**, mis-proportioned as compared to real life

Emilie pauses and concludes if it does not contain these then it is not a good painting. Emilie has in effect created a data model to define good paintings but mistakenly fit her "model" *only* to the style of Van Gogh's work. The model is very good when defining a Van Gogh as a good painting, but the problem is it is *only* Van Gogh's paintings that fit the definition. Imagine Yvonne taking this list of attributes and applying it to other works of art. Leonardo Da Vinci's *Mona Lisa* would not be considered a good painting with the absence of bright colors, textures, swirls, and short strokes of paint. Not to mention its realism. No, by Emilie's definition it is not a good painting at all. Nor would *Washington Crossing the Delaware*, the *Last Supper*, *American Gothic*, a painted circular plate, or any famous painted portrait be considered a good painting. The data model may work for other impressionist paintings, but its use by this definition is limited. It is overfitted. An overfitted model is a prime source for a data dupe.

When a model is overfit, it is *overdescribing* the attributes from its reference data set. The creator of the model may exude confidence that it is an excellent model because of how well it fits the data. But if it fits too well, then you should be skeptical about its usefulness beyond its reference training data set. The very next question to ask when presented with a "strong" model is how well it performs against a random sample of other data. In our example, Yvonne could randomly pick objects from another gallery and apply the model to soon learn that it was not very good at defining a good painting. A data model that is overfit is another example of "too good to be true."

FACT VS. OPINION DATA

In grade school, you may have been fortunate to have a teacher who guided you on the mechanical differences between writing facts compared to writing opinions. There certainly are a few key elements that differentiate each, especially when the writer's intentions are clear ("in my opinion . . ." or "It is a fact. . . ."). However, writers are not always clear nor concise and the ambiguity between the two can create deception, where opinions masquerade as facts. First, let's start with some definitions.

A fact is a thing that is known or can be proven to be true. An opinion is a judgment, an idea, or a concept that can neither be proven nor disproven. An opinion is sometimes a personal belief, an expression of one's feelings and thoughts, or a hypothesis. Although at that point a hypothesis starts to encroach on facts since a hypothesis can be tested to be true or false. Therein lies some of the difficulties of separating fact and opinion—they can and often are mixed together.

In the days of print newspapers, finding opinions was simple. There were clearly marked sections for them separate from "The News," "Local News," "Sports," and so on and clearly labeled "Opinions" or "Editorials." The guideposts to know what was fact and opinion were well defined and some would say the two rarely crossed paths, except maybe the sports section writers editorializing the performance of their favorite weekend sports teams in Monday's edition.

In the digital era, things of course got a little more difficult to label. Newspapers transformed into online versions of themselves and for the most part continue their tradition of sections dedicated to news facts and opinions. Stories appearing on their digital "front pages" get tagged with the word "Opinion" when appropriate. Other digital-native news sources are not bound to the same type of tradition. The *Huffington Post* for example does not offer a distinct Opinions section although many original articles are opinion pieces, leaving the reader to discern for themselves between opinion and fact. Likewise, fact and opinion stories are not differentiated on Facebook when stories breeze through your newsfeed. Although most people (41 percent) prefer to get their news from television, a growing number are seeking news from online sources (37 percent). Online news sources are increasing among social media users, with 15 percent preferring them as a resource. Further mixing up fact and opinion occurs when traditional news sources are clipped and quoted on social media, leaving out important context. So, what is the best strategy to avoid being data duped when sifting through opinions and facts? Start with the data.

Like any good data scientist, a great place to start is to look for data within the piece you are reading and assume everything is an opinion unless there is evidence to show it is factual. Remember the definition of a fact is that it can be proven to be true, so without data, it is doubtful it is a fact. Wikipedia is a rich resource for facts such as this example:

- **Fact**: "The Industrial Revolution, now also known as the First Industrial Revolution, was the transition to new manufacturing processes in Europe and the United States, in the period from about 1760 to sometime between 1820 and 1840. Mechanized cotton spinning powered by steam or water increased the output of a worker by a factor of around 500. The power loom increased the output of a worker by a factor of over 40."—Industrial Revolution. en.wikipedia.org/wiki/Industrial_Revolution
- **Opinion**: The Camry is a great midsize car that hits all the marks. It boasts nimble handling, a comfortable ride, two strong engine choices, an upscale interior, and a user-friendly infotainment system. Should we Buy the Toyota Camry? Yes, the 2020 Toyota Camry is definitely a worthwhile purchase. The Camry may be on the pricey end of the midsize car spectrum, but this well-rounded sedan is worth the money.—Toyota Camry Review. *US News.* 2020. cars.usnews.com/cars-trucks/toyota/camry/2020
- **Mixed Facts and Opinion**: The whole Camry lineup performed extremely well in the EPA's tests, and a four-cylinder model did brilliantly in our real-world highway testing. The hybrid's upper trims sacrifice some efficiency at the altar of luxury, however. A four-cylinder Camry SE returned 45 mpg in our highway test, beating its own EPA rating by 6 mpg—and making it the most economical nonhybrid car we've ever tested.—2021 Toyota Camry. *Car and Driver* 2020. www.caranddriver.com/toyota/camry

In the fact example, there are references to proper nouns (Europe, United States), specific dates (1760, 1820, 1840), and numbers to support the claims of increased output (500, 40). Clearly, this is factual and there is no suggestion of judgment or feelings about the topic.

In the next example of opinion in the form of a product review for a Toyota Camry, we can read the feeling of the writer, who is certainly advocating for this car, as it closes with "[the] sedan is worth the money." Without the facts about how much it costs, how do we know it is *worth* the money? There are no data, no formula to justify its worth. Worth is a bit of an ambiguous word and maybe what is worth it for someone could be different for others.

The last example is a slightly different review that includes a mix of both opinion and facts. It provides a reference to specific tests and numbers proving their results.

These examples may look too simple, and for the sake of demonstration, they are easy to discern between fact and opinion. However, we are confident that in media there are many more complex examples, and the ones most difficult to separate between fact and opinion are those that mix the two. That is, an opinion propped up with facts. This loose and sometimes biased interpretation of facts is the heart of the type of data duping we are seeking to guide you against. The use of numbers creates the guise of fact but is driven by motive, influence, and judgment that is often opinion with data, and in our opinion this is not a good use of data.

In the spirit of being good data guides, here is a checklist of what to look for when deciding between facts and opinion:

✓ First, approach everything as if it is an opinion. If it is not, soon the facts will reveal themselves.
✓ Specifics: Like a good reporter, is there a reference to person, place, or things?
✓ Numbers: Are there details such as numbers, dollars, and percentages?
✓ Dates and times: When? Are there specific timelines when appropriate?
✓ References: Was it the best baseball team ever, or in the moment did it just feel like they were having the best season? Why? Does the writer give references to back up their statements?

Having discussed the difference between fact and opinion, how do you think you would do in a test of a random paragraph? Once armed with a little data defense, determining fact from opinion appears more clearly. The data dupes are more likely to come when you are not explicitly looking for it or it is presented out of context. Scrolling through your newsfeed and browsing social media may be when you are most vulnerable, but it could also occur at work or in other interactions where you have less rather than more information. This all made us curious. If using the guidelines above help people read the difference between fact and opinion, could a computer be trained to do the same?

We evaluated a few approaches to classify written text into facts and opinions. The results are surprisingly good. When the computer is "trained" with data similar to the above from Wikipedia (facts) and product reviews (opinions), most of our computer data models can classify a sentence as being more fact than opinion.

A note of caution. Although computer models can successfully classify sentences between those appearing to be fact and those as opinion, we really need to understand that the computer models were putting weight on the *structure* of the sentences and not the content. Those deemed to be facts tend to have more specific references—actual people, places, and things—and specific details to dates and events. Some of this may be an artifact of the training source, which creates a bias. Also notable is that these computer models were trained exclusively from a single source and may have been honing in on the particle writing style of Wikipedia, rather than the faithfulness of the contents. In other words, if a sentence looks like it came from Wikipedia and we know Wikipedia is a trusted source of facts, then the model deems it a fact. This may not always be true and both are examples of training bias and how models can be misleading. The computer is not like the sci-fi computers of the future seen in movies. It is not *actually* reading the facts and cross-referencing them to be true, it is only looking at how the sentence presents the information. This means that a nonfactual sentence could be written to appear as fact when it is not. The sentence "George Washington was the first US president from 1789 to 1797" would rank as fact just as "George W. Bush was the first US president from 1879 to 1977." The facts themselves are not checked. Opinions certainly have a type of style in their writing, and that is all we get from this algorithmic approach. Sentences are judged to be fact or opinion, yet a reader needs to consider the whole article to make better judgments. An article with a mixture of factual statements and judgmental opinions makes the task more difficult, and again this is often where you will encounter a data dupe. The numbers and data give legitimacy to an opinion and can be misused. Nothing beats a real person reading an article for substance, context, and accuracy. At least not yet.

Can a predictive model read a paragraph and determine (predict) which sentences are fact-based and which are opinion? Can it be more consistent than people? Yes, we believe so. Consistency is probably the key difference between computers and people. People, due to personal bias, distractions (such as when people are absently scrolling through their newsfeeds), and changing information, are prone to inconsistency in identifying facts and opinions. This is where a predictive data model can help you not to be data duped![15]

WHEN MORE IS LESS (USEFUL)

With the internet age, we have more information available at our fingertips than ever before. We have the ability to know almost everything known or knowable, nearly instantly and nearly for free. Information is readily available to view and inspect and prod and cajole and share and do whatever we want with it. And there might be part of the data duped problem.

We started this chapter with an example of how the military confronts data deception as a means of fighting a battle. Not a physical battle, but one of information. It is an apt analogy for data deception in general, one side opposing the other. This may be among the local debate teams, decisions you make at work, or spirited discussions with your uncle Dilbert during the holidays.

In the fifth century BC, Sun Tzu was credited with writing in the *Art of War* that "all warfare is based on deception." He meant it in more practical terms about where and when to place your army, but there is a lesson in his words for us as well. The lesson for people working with data and those of us trying to understand it is the need to know that, with or without intention, data can be deceptive. It can be used to misrepresent and misguide. The war may be for a particular decision (how much to save for retirement) or simply to garner your attention (news alert—"this threat could be in your neighborhood"). Were General Sun Tzu alive today, he certainly would make use of the internet. He likely would use it to slow down his enemy's advance by overwhelming them not with weapons, but information. Yes, data can stop an army in its tracks. Nothing confounds a decision better than information overload.

There is a lot of data in the world and some of it is even useful. By many accounts, there are 40 to 45 zettabytes of data in existence and its growth continues exponentially, driven by more digital tools from phones and smartwatches to IoT devices.[16] A zettabyte is a lot of data, specifically 10^{21} or 1 followed by 21 zeros of data bytes. A byte is eight bits of 1's or zeros, which is how numbers are stored on computers. A single typewritten character "D" assigned to the number 68 requires one byte and is stored as a sequence of bits as 01000100. Consider that the complete works of Shakespeare can be stored in five megabytes (5 × 1000^2 bytes), and the amount of data that exists in the world is mind-boggling.[17] In more practical terms, the data duping you are likely to encounter will come from processed and summarized data such as in online news. Still, there are billions of those sources.

The problem is with all those sources, it may be difficult to sort out the truth from misinformation, or at least the varying opinions. Try searching for "How much should I save for retirement" and you get *millions* of answers (we actually got 56,800,000, but you get the point). There is a lot of information out there, but perhaps this is because there is disagreement on how much money someone may need in retirement, or the various ways one could save, or the products that are best to invest in, and on and on. Alternately, maybe there is a lot of agreement on this question and there is just a lot of duplication between the sites offering answers. Perhaps with a mixture of ways to answer a seemingly simple question between opinions ("stocks are better than mutual funds!") and the facts of the math (i.e., the future value of $10,000 at 6 percent interest), makes for many solutions.

Let's take something we could all easily agree upon—"the rate of gravity." A quick internet search results in . . . 3.5 million web pages! Wow! Perhaps we erred in the way we phrased the search since in addition to the rate of gravity on Earth (9.81 meters^2) we also found the rate of gravity for Mars (3.711 meters^2) and some other "search-noise" with references to movies named Gravity and measures for Specific Gravity (the amount of solid in fluid) for batteries and saltwater fish tanks. This detour of web searches shows us it is easy to find information, but more difficult sometimes to find exactly the information you need. Verifying a fact takes time, and an abundance of information makes the task even more challenging.

In 1970, author Alvin Toffler first coined the phrase "information overload" in his book *Future Shock* about the impacts on people of a rapidly changing world. He described it as "the distress, both physical and psychological, that arises from an overload of the human organism's physical adaptive systems and its decision-making processes."[18] At the time he had not yet witnessed the internet and the plethora of news, articles, social media, and just pure data it displays. The difficulty we have sorting through all this online content can contribute to information overload and make us ineffective in our ability to process information and make decisions.[19]

With more data and more information, can we say people are smarter, more informed, and more knowledgeable? How could we test this? We must be getting smarter, right? If so, we would expect to see higher education scores. The Flynn Effect has measured an increase in IQ scores over time rising more than 10 points since the 1990s when the internet began to grow in popularity. Ten points may not sound significant, but given the typical IQ test score is 100 and genius level is 150, it is quite an achievement. However, most experts agree the Flynn Effect is more a result of better health and nutrition, more education, and improving worldwide standards of living[20] [21]

And with all this information, perhaps there are fewer conspiracy theories, although the internet might be contributing to these. Uscinski and Parent write in their book *American Conspiracy Theories* that the number of conspiracy theories over time is relatively the same, they just spread faster in the internet era than they did before.[22]

There is a paradox. At a time when more information is available, more people seem unaware of the difference between truth and untruths. The problem is not that the information is absent, it is finding the truth is more difficult than ever. Sometimes depending on the topic, it is hidden under the troves of misinformation.

KEY POINTS

- Half Right: If the information you are getting or the estimate you are making is only half right, will this change your decision one way or the other?
- Correlation is *not* causation. When two things seem to be related, it could be 1) a result of a third influencing factor or 2) coincidence.
- Mean vs. Median. Skewed data can "tug" on the mean average in the direction of the skew, leading to misleading conclusions about the "typical" thing being measured. When presented with a claim about the average, consider 1) what the distribution of the data is, and 2) if you suspect it is skewed, how much could it be influencing the average?
- An algorithm or data model that claims to be highly predictive may be a result of over-fitting to the training data set. When a model overdescribes the data, it may lack usefulness when applied to other data. Ask: How well does it work on a random data set? Remain skeptical of too-good-to-be-true models.
- The difference between fact and opinion is simply a fact can be proven to be true or untrue, while an opinion is an unprovable judgment. Fact and opinion statements are often mixed. The facts of one person's statement bring legitimacy to their opinion points of view. Learn to separate the two and not assume some facts prove an opinion. Assume everything is an opinion until facts (data) are presented. Ask: Where are the specific data points?
- More information is not always better. Excess information and data can result in information overload, which can limit your ability to make effective decisions. The growth in the availability of information also means misinformation can enter the decision process, leading you to being data duped. Misinformation presented with data can appear credible. Ask: What is the source?

4

Data Duped in Media

Reality Checks for News and Marketing

Ma·lar·key /mə'lärkē/ Noun
Meaningless talk; Nonsense. Unknown origin. Popular among Irish
Americans in the 1920s. Example: "don't give me that malarkey!"

—Another Definition of Nonsense, Oxford Dictionary

MEDIA AND THE MALARKEY METER

"What a bunch of malarkey!" Growing up you might have heard this
expression used by your grandparents in response to some nonsense
or misinformation. A ruffle of the newspaper, a sour-looking face, and
then . . . "Malarkey!" The way we gathered our news in the past, in
old-fashioned newspapers, on nightly TV news, or just gossiping with
our neighbors, certainly has changed over time. Those ways have been
replaced with Facebook, YouTube, Snapchat, and others. What has not
changed is our reaction, and that is precisely what news organizations,
advertisers, and sometimes your neighbors, now using social media, are
trying to create—reactions. Reactions are better than indifference, and
if the local newspaper could get you to ball up the paper or entice you
to yell at the TV newscaster, why, that would be all the better.

Today, like then, getting a reaction is part of the business. This leads
us to the malarkey meter, a bit of a self-test on the trustworthiness of
news, marketing, or anything else you might read or see in media. One
place to start judging information might be as simple as this test. How
can you know if the information is good? On one extreme some things

have a high level of trust and we think of these things as true. The number of feet in a mile and the tallest mountain on Earth are both known facts. At the other end of the malarkey meter are dubious statements, misleading claims, and data deception. So egregious we can only call them "Malarkey." The term, adopted from European immigrants, was very popular in the early 1900s but quickly faded by the mid-century. However, its uses, according to Google trends, has been increasing since the early 2000s. Perhaps there is a rise in the amount of malarkey people see in media.

In the previous chapters, we provided some tools for data defense and strategies to help you navigate disingenuous data. In this chapter, we will show how data can be misused to influence an audience or just to grab their attention. We will see a range from subtle deception to the outrageous and along the way call out the "malarkey meter" when we see a data dupe.

NEWS

Long History of Fake News

Malarkey, fake news, and distrust of the media are not new. The notion that a free press and more access to information can lead to more truth and knowledge is a noble ideal. It is one shared throughout history from the early inventors of the printing press to modern times. However, even John Adams, a member of the Founding Fathers and the second US president, had doubts. At the time he was helping shape our young nation and put forth ideas of liberty and freedom, he turned to read works of others, such as French philosopher Nicolas de Condorcet. Condorcet, writing in 1794 while imprisoned by the French revolutionists, advocated for many ideas that would seem modern, such as equal rights and free education. On the topic of free press and truth, he wrote:

> It is even apparent, that, from the general laws of the development of our faculties, certain prejudices must necessarily spring up in each stage of . . . [mankind's] progress, and extend their seductive influence beyond that stage; because men retain the errors of their infancy, their country, and the age in which they live, long after the truths necessary to the removal of those errors are acknowledged. (Condorcet from *Outlines of an Historical View of the Progress of the Human Mind,* 1795)[1]

Condorcet believed the solution to misinformation—the eighteenth-century equivalent to data duped—was more access to knowledge. Not just for the aristocrats but for the commoners too. The more information that could be shared, in particular by a free press, then the less likely untruths would be spread, "because men retain the errors . . . long after the truths necessary to [their] removal . . . are acknowledged." John Adams seemed to agree with the idea that mistruths, once established, are difficult to overturn. Upon reading this passage in his copy of Condorcet's book, he wrote in the margins: "There has been more new error propagated by the press in the last ten years than in a hundred years before 1798." In other words, as the young nation was embracing its newfound freedoms, including the ability to publish newspapers without license and authority by the government,[2] Adams was recognizing this noble idea of truth in the news was partly failing. But why?

One thing the feisty Americans were learning was inspired by Francis Bacon in 1597, that "knowledge is power," and they used the power of the news in printed pamphlets to organize and build support for the American Revolution. They also used it, as Adams's musings suggest, after the revolution to take aim at their elected officials. A familiar trend that persists today. George Washington, too, noted as he was leaving office that he was growing tired of being "buffeted in the public prints by a set of infamous scribblers."[3]

Although criticizing government leaders is not necessarily in the same category as misinformation, the young press did not waste time learning how to stretch the truth and manipulate readers for their purposes. In 1765, Britain imposed an unpopular tax on the colonies called the Stamp Act. Colonist Samuel Adams leveraged the press by writing in the *Boston Gazette* about his distaste for the tax and, in particular, attacked the local British representative, Royal Lt. Governor Thomas Hutchinson, writing several untruths about him. The response? The readers without the ability to fact-check Adams' claims grew enraged and stormed Hutchinson's residence. They became a mob and ripped apart the home and took everything possible. As historian Eric Burns wrote, "they were old men, young men, and boys barely old enough to read, all of them jacked up on ninety-proof Sam Adams prose."[4] The irony is that despite what was written about him, Hutchinson *did not* agree with the Stamp Act and the collection of tax. The news had misinformed and led to actions aimed at the wrong person.

This may be an extreme example of how our actions can be influenced by misleading information in the news, but it is not isolated. We make decisions all the time based on what we read, hear, and see in the news. The weather report helps us decide to take an umbrella. The latest economic forecast might determine how and when we invest . . . or spend money on a vacation. The news, its facts, and its data are important, and avoiding being data duped can help you be more informed and may even help with all those decisions.

Decline of Traditional Media and Rise of the "Free" Press

News has become more complicated with the decline of traditional news media and the growth of social media. The number of newspapers grew substantially in the early days of the United States, peaking in the early 1900s with about 17,000 different publications. Circulation continued to grow through the 1990s, when online media began to capture part of the news audience. Circulation dramatically began to decline in the mid-2000s, coinciding with the growth in popularity of Facebook, Twitter, and other similar sites. Facebook launched in 2004 and was made more widely available to the public in 2006. By 2010 they acquired more than 500 million users. They would double that number to 1 billion by 2012. By 2020, they had 2.7 billion users, and although these are not all people accounts, most of them are, and it represents a sizable portion of the 7.5 billion people in the world. The migration to social media likely began as expected with users seeking another "social" outlet rather than a news source. However, social media quickly learned the more they could get their users to connect to their apps the more successful they would become.

The invention of Facebook's Newsfeed took shape not only to show you the news of your friends and family, but also the news of the nation and world. Engaging users for a longer period of time became a goal for social media, and presenting news was just a part of that process.[5] The purpose of engaging users longer was driven, quite simply, by advertising revenue. By captivating their users' online social media, these websites can extract more information about their users' likes and preferences in addition to their demographics. The result is the ability to successfully "micro-market" advertising to each user.

Think about that for a moment. In the traditional approach in advertising where *Mad Men* meets math, marketers would divide their potential customers into segments. These segments were based on data, at least as much as could be gathered about them, and using these data, a creative marketing strategy and a message was developed to target each group. In the earlier days of marketing, these segments and their messaging were broad—after all, whatever they wrote and put on a billboard or a television ad had to resonate with a lot of people and was displayed for several weeks at a time. Broad messages with few specifics. Newspaper headlines worked the same way. Most newspapers were published daily, while a few others were weekly. The opportunity to get their messaging right, with the right stories and headlines to attract the most readers, was limited by those publication cycles. Newspapers often dedicated 60 percent of their printed space to advertising; showing readers eye-grabbing headlines with a story snippet followed by "continued on page x" was part of the art of engaging their readers longer and showing them more advertisements. The task of matching advertisements to readers on the same page was thoughtful targeted marketing. The ads for hunting equipment in the sports section was by design, as was life insurance in the obituary section. The problem with this traditional approach is how often it failed. Most people reading the obituaries may not need life insurance, because they already have it. Those looking for the latest scores for last night's games might not be hunters. Both segments of readers still got a printed advertisement for life insurance and hunting equipment, along with every other person who bought a paper. Most of the printed ads would go unseen, and an ad that is not shown to anyone is a wasted expense.

In the marketing world, there is endless debate on the effectiveness of various types of advertising, captured in a term called "lift." The concept of "lift" is how many more people purchased the product after seeing the ad compared to before. The amount of lift in terms of sales revenue less the cost of the advertising is the profit gained from creating the ad. In short, spending money on ads is supposed to make more money than not spending the money. Again, there is lots of debate about when and how people respond to what they read and how this affects their consumer behavior. Sometimes people just buy life insurance and the timing of those ads is irrelevant. Of course, the combining of social media, advertising, and news in one compartmentalized space in our lives dramatically changed all of this. As a society we have been cutting our cable cords, tuning out our television news stations, and canceling our newspaper subscriptions.

For Facebook and others, this has been tremendously profitable. In 2020, Facebook's revenue grew 33 percent over the previous year, generating $28 billion, of which 96.8 percent came from advertising revenues.

Engaging social media users is about more than just news, but the news is an important part of the strategy. A Pew study revealed that 52 percent of Facebook users get their news on the site.[6] And with more people using digital news sources, the decline in traditional newspaper revenues is in a downward spiral. The shift in our behavior has an important impact on the integrity and perhaps even our very definition of what we call the news. Before the internet and the development of social media, there were barriers to how news and information were distributed, namely, it was expensive. Although as Samuel Adams demonstrated with his attack on Hutchinson, it was possible in the past to fabricate misinformation, but it was limited to a few influential individuals. In the 1760s there were only a handful of newspapers in the United States and *only* 17,000 at their peak in the early 1900s. By contrast in the social media era potentially any user with a small bit of sophistication can create a news story or at least one that *looks* newsworthy and publish it to a broad audience. From blogs to Instagram, to newyorktimes.com, the number of places we can get the news is expanding. And with the cost of news distribution nearly eliminated, a consequence sometimes is the truth.

Perhaps our susceptibility to being data duped in the news is because of the rapid change of the medium from traditional and dependable news reporting toward social media news. But again, traditional news was not always immune to data deception. Newspapers of the past and present are still motivated by eye-catching headlines that can help drive revenues and, as a result, have their share of misinformation. However, given the expense of producing a newspaper generally, in the past there was more of a focus on accuracy and avoiding deception. With this traditional approach, readers gained trust in the medium and generally did not need the data-duping defense skills we are advocating. As a result, people tend to believe the news from traditional news sources and their new social formats alike. An Ipso MORI study revealed that 46 percent of Americans admitted to believing a news story before later finding out it was fake, despite the fact 63 percent *believed* they could spot a fake news story.[7] Studies like these and other observations of deliberate data deception, misinformation, and propaganda have caught the attention of educators.

In 2014, Finland began integrating media literacy programs into their school's curriculum beginning with the earliest grades. The goal is to teach children to evaluate truth versus misinformation, in particular in online sources.[8] As digital natives, you might believe younger people are less vulnerable to online fakery, but a Stanford University study showed students in middle school, high school, and college struggled to identify misinformation. They wrote, "when it comes to evaluating information that flows through social media channels, they are easily duped." Whether you are young or old, educated or street-smart, there continues to be a need to avoid data duping.

The Cost of Trust

With mainstream news media and social media actively vying for your attention, which of these can you trust? With questionable intentions toward truth and data accuracy, spurred by conflicting motivations— Facebook wants to grow engagement and users, and the *Washington Post* writes, "Democracy Dies in Darkness"—is there a cost of trust?[9]

Among the motivations for writing this book was the realization that the environment in which we all consume data is rapidly changing. The flow of data in social media, marketing, and the news are just examples of how information is seeping into many parts of our lives in ways that did not exist just a few years ago. Are you more likely to be data duped now compared to the past? Probably so, not only with this influx of data but also because the mediums where we consume it are changing too. Like all other paradigm shifts in history, certain trade-offs occur at the onset of a "revolution." During the Industrial Revolution, the change impacted workers, requiring them to rapidly acquire new skills and adapt to new ways of manufacturing. There were trade-offs and costs—some workers found they were unemployed, others discovered they could now afford items made more cheaply. There was the good and the bad. Likewise, with data, information, and its transformation into news, it requires some costs to our society, especially for those unprepared. The transition in how we consume news has consequences. One example comes from the demise of local newspapers.

One thing we learn from the decline of traditional media and its replacement with free social media is a decline in the inspection of the truth. Local newspapers are more relied upon as a source of local events from topics of crime and community to government affairs.[10] As it relates to government, local newspapers and their journalists play more of a "watchdog" role than any other source. They are the reporters who attend local board hearings and ask questions about how their government is responsibly making decisions.

Researchers found the closure of local newspapers did not create suitable "watchdog" substitutes in social media news. They measured the local government's borrowing costs, usually through issuing municipal bonds, increased by 5 to 11 basis points. Although this is a small fraction of a percent, the average bond issues result in an increase of $650,000 per issuance. They also found other costs related to increasing spending and growing budgets. The watchdog effect and the inspection of the truth when social media news replaces other sources are absent. There is purpose in those investigative reports and attention to the truth. The research shows traditional news sources have value, but sometimes with all of those reports from the city council meetings, well, it might just be a bit too boring.

Provoking the Truth

As we cover the spectrum of new sources and consider the costs and trade-offs of the "free" press, we come back to the economics of news, namely how to generate revenue from advertising and subscriptions among a crowded field of news sources. When such a large number of people are online daily using their social media apps, how do you convince them to pay to get (apparently) better news from another source? The answer might be to look beyond the boring and mundane news and consider how effective those other sources can be at provoking their audience to buy their offering.

Misleading and exaggerated headlines sound like obvious data duping. It is also the type of thing that will stop a reader in the middle of their finger-scroll. What can make them successful in engaging readers is when, on the surface, they do not appear misleading. Obvious deception will more often be overlooked: "Aliens Land in Washington, Invite President to Fly around the Moon," would barely get a read, a like, or a reshare on social media. It is humorous but obviously not true. Instead, headlines with some believability and exaggeration often do better with readers, such as: "Millions Plan to Scale Fences of Area 51: Home of Government Secret Alien Research." Area 51 is a real US Air Force facility that was used secretively in the 1950s and 1960s to develop reconnaissance spy planes. The government's denial of activities there for the sake of national security led to many conspiracy theories about space aliens. Part of the headline has some truth—Area 51 is a real place—while the rest of it provokes the reader to want to learn more.[11]

Our brains react more to novel stories and unexpected headlines. We are drawn to the surprising, sometimes outlandish headlines because a routine headline is more expected. In other words, routine news stories do not grab our attention. If we read, "Weather at the Beach Will Be an Average 75 Degrees This Weekend," it would not seem interesting. Because of course, your expectation is the weather at the beach will be nice and sunny. In contrast, "Plane Kills Beach Jogger in Emergency Landing" is more likely to invoke a response and something you would share.[12] The story is unexpected and improbable and according to researchers triggers a neurological response in your brain's hippocampus that lights up sensors visible on an active MRI scan.[13] In other words, a shocking and surprising headline that is within the boundaries of believability can really get your brain juices flowing. People are more likely to share these stories. These are also prime opportunities to be data duped. A story with just enough data can be seen as more credible. If someone pushes the envelope a bit more with data, it may even seem shocking. A story of climate change impacting our daily lives in 40 years might be commonly accepted. But, although a news story indicating climate change was accelerating and would change how we live in just seven years might be intent on misleading readers, it likely will also garner a lot of attention and be reshared. In this way, data duping can propagate to others.

A study by MIT researchers of 10 years of Twitter posts showed retweets of false and misleading information reached more people and occurred six times faster than truthful tweets. They conclude Twitter users prefer to share falsehoods over the truth.[14] The intention of the Twitter users is likely not to purposely deceive but to quickly share something that seems unimaginable, yet believable, perhaps even useful to those they are sharing with since the data are stark and surprising. Interestingly, the study also controlled for the promotion of tweets by Twitter's automated resharing tools. They found the automated tools were more neutral than the humans, equally distributing truthful and false information. To the tools, the news was the news. Where they had no emotional connection—typical—we humans were lighting up our brains with all of the shocking data!

The examination of online social media behavior goes a long way to explain how misinformation and deceptive data can "grow legs" and run around the globe faster and to more people than the truth. Provocative information can be more intriguing to its readers. Amid social media grabbing people's attention, news agencies find they need to do the same to engage their readers. Newspapers and television news, the traditional media, have come to form a wide spectrum of provocativeness from the absurdness of grocery store tabloids to award-winning journalism.

This Just In . . .

"Grocery Prices Soar the Most in a Decade in 2020."

With this eye-catching local TV news headline from March 2021, what is the first thing that comes to mind? Are food prices out of control and will we have enough money to afford to eat? This might be the type of reaction the article is seeking to provoke. As we read the article, however, it may not be as dire as the author indicates. Let's take a closer look.

First, at nearly any point in time, grocery prices are "the highest they have ever been." This is natural because like all commodities their prices tend to rise with inflation. Negative inflation does occur although it is rare for groceries. Even in the post-2008 Great Recession, food prices still increased. The smallest was 0.8 percent in 2008.

What this means is food prices in 2020 are certainly higher than in 2015 and 2010 and so on. So, the exclamation that the current food prices are at an "all-time high" really should not be too surprising, unless someone has found out how to grow a cheaper tomato.

Next, the article's mention of grocery prices *soaring* by 3.4 percent makes us curious about how significant the financial impact of

3.4 percent is on the typical household. A little investigation from the Bureau of Labor Statistics shows the typical (average) household spends $4,636 annually on groceries or about $90 per week. Therefore, a 3.4 percent increase would amount to about $3 more per week. Given the 20-year average annual change is 2.0 percent, the typical household would already be expecting to pay about $1.80 more per week. While 3.4 percent is numerically much larger than the 2.0 percent long-term average, in practical terms it means about $1.20 more per week or about $60 more per year. Certainly, 2020 was an exceptional year and it tested food supply chains in many ways, some of which caused an increase in producing and delivering food to grocery stores. Another factor may also be related to the fact that more people were eating in (buying groceries) rather than dining out. An increase in demand for groceries could influence costs. Although in 2020, there was a similar increase in the cost of dining out (known as food-away costs), which we believe is related to the costs of COVID protocols, there have been off-setting trends in the past between food-away costs and grocery costs. In 2016, for example, grocery costs decreased 1.3 percent, an uncommon event, while dining-out costs increased 2.6 percent. Nothing, however, compares to the period from 1974 to 1981, when *average* annual groceries increased by 8.5 percent each year. Perhaps in 1974 when prices increased by 14.9 percent in one year alone, then the headline "Grocery Prices Soar!" might have been a better fit. The story in 2020 is a bit less dramatic by comparison. Although the data are factually accurate, the intent is more dubious. In other words, it's a ton of malarkey.

Data Duped Guide to Truth in News

At this point, it might be useful to categorize the types of misleading content and where you are most likely to encounter data duping. The term "fake news" has grown in use in the past few years and at times has taken on different meanings, such as when politicians refer to "fake news," which is not a misstatement of facts, but rather an unfavorable framing of the politician. In other words, when a politician uses the term "fake news," they may just be saying they do not like how it portrays them. In a recent study, researchers helped define categories of fake news based both on its truthfulness and the intent of the authors.[15] Similarly, we can construct a method to evaluate news and data duping based on the author's intent to deceive and the method of communication.

Data duping in the media can occur across a wide span of mediums. We will not name names, but you can quickly get a sense of which publications might fit into these categories based on the accolades given to their authors. On the far end of the truth scale is science. If your data source includes a few who have been awarded a Nobel Prize in science, then this is a good sign. In the true News category, you may find a few Pulitzer Prizes in Journalism. Infotainment is a little more difficult to identify, since by definition these media sources are branding themselves as a type of news but without the same standards of truth. A data source in this category may be best characterized as a grocery store tabloid and, if they have been sued and lost to a celebrity, well you might want to stay away from relying on them as a data source. Finally, there is entertainment media, and these, well, should be obvious. They are not intended to be factual. If recognized with an Emmy or Academy Award, then you know this media is meant to entertain more than inform. (See figure 4.1.)

MARKETING

"New and Improved!" and other useful marketing phrases once were the standard for advertisers when print and in-store displays dominated the marketing message. Now such puffery about the merits of a product's features is easily dismissed as consumers become discerning.

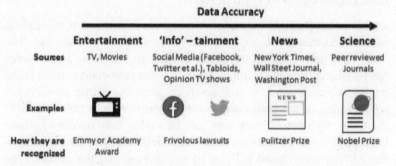

	Data Accuracy ➞			
	Entertainment	**'Info' – tainment**	**News**	**Science**
Sources	TV, Movies	Social Media (Facebook, Twitter et al.), Tabloids, Opinion TV shows	New York Times, Wall Steet Journal, Washington Post	Peer reviewed Journals
Examples				
How they are recognized	Emmy or Academy Award	Frivolous lawsuits	Pulitzer Prize	Nobel Prize

Figure 4.1.   Knowing the source of information can inform its data accuracy.

In the United States, marketing and advertising are regulated by the Federal Trade Commission (FTC). The FTC does not review and approve every advertisement but they do create guidelines and regulations about fair marketing and have the authority to enforce those rules, sometimes with hefty fines. In 2019, the FTC levied a $5 billion penalty against Facebook for its deceptive claims about how it was protecting users' private data. Stemming from earlier complaints, Facebook marketed the ability for users to protect the sharing of their personal data through user-controlled privacy settings. However, the FTC discovered there was a loophole. Unknown to users, when their Facebook friends installed other third-party applications, Facebook's agreements allowed those applications to override the settings and further gather personal data from those users' friends, independent of their privacy settings. In most cases, the individuals had no indication their data were being taken, despite the settings they made on their Facebook accounts. As the FTC wrote, "Facebook subverted users' privacy choices to serve its own business interests."[16]

Another large fine for deceptive marketing was placed against automaker Volkswagen in 2016. Although the fine was smaller than Facebook's, the overall costs of their deception was greater because, in addition to fines, the FTC required Volkswagen to make restitution to their customers by offering buybacks of their vehicles and canceling lease agreements. The deceptive claim was based on data and an incredible claim in their marketing commercials. Volkswagen advertised their "clean diesel" engines reduced nitrogen oxide emissions by 90 percent and further had fewer emissions than similar gasoline engines in cars.[17] Impressively, Volkswagen had data to support this claim from emissions testing. Typically, the testing involves connecting a vehicle's exhaust to an emissions measurement device and testing the car while it is running in a lab-like setting. The problem? Volkswagen adjusted its software controlling emissions such that the emissions controls only reduced the nitrogen oxides *while* the car was being tested. Out on the road in everyday use, those emissions controls were not functioning. They had schemed to both deceive the testing facilities and then use the poorly collected data in their advertising. In the end, Volkswagen was required to recall at least 85 percent of their impacted vehicles for a total estimated cost of $14.7 billion. To give context to the severity of their deception, in 2016 before the fines, Volkswagen's total net income was $3.35 billion.

Your Mileage Will Vary

In general, the FTC has broad guidelines for marketing along the lines of thou shall not intend to deceive, misrepresent, or mislead with data. And if you have data and use phrases like "experts recommend" or "studies show," then advertisers need to be prepared to show it. The FTC also routinely reviews guidelines, eliminating those that are no longer needed and adding new ones as the times change. For example, apparently in the 1950s advertising nuclear fallout shelters was a tricky business, and the FTC had to define specific guidance on what determined a "fallout shelter." To be advertised as a shelter it needed to withstand at least 40 times the amount of radiation of observing a blast in person and there were detailed formulas for blast protection based on explosive pressure—2 miles from a 5-megaton bomb and 10 miles from a 100-megaton bomb were common references in the regulation.[18] Much like how today we choose refrigerators based on the data of the energy efficiency guide labels, we can imagine the families of the 1950s comparing fallout shelter labeling.

Fallout shelters and home appliances demonstrate sometimes the math and methodologies can get complicated for consumers. For deliberate data dupers, it creates opportunity for advertisers to unfairly portray their products. In these situations, the FTC has some very specific guidance, such as new vehicle mileage ratings. Advertisements for new cars often tout fuel efficiency as the main benefit, although typically fuel costs are only 16 percent of an owner's expenses, while the car payment is 60 percent of monthly expenses.[19] The data-duping lesson here is that marketing is overemphasizing the benefits of a less relevant feature while downplaying the most important factor: the purchase price.

To maintain fairness in fuel economy advertising the FTC creates standards for advertising, and if you have looked at a new car you will notice the large window display of the fuel economy for both highway and city driving. For the technical part, these data are gathered in a lab while a driver puts the vehicle through simulated driving conditions and a computer measures the fuel consumption. Standardization helps provide data objectively, but one thing to consider is the fact that it is measured in a *lab setting* and simulates a driver's typical behavior. What if you are not a typical driver? Perhaps you live in a busy congested city where you spend hours stuck in traffic and needlessly burning fuel while completing your daily commute. Maybe you are an aspiring NASCAR driver or drive to the grocery store like you are on the wheel of a firetruck. Aggressive driving—fast accelerations, quick stops, and avoiding using cruise control on the highway—can have meaningful impacts on fuel efficiency, up to 38 percent.[20] Further, the temperature has a strong effect on fuel efficiency. If you live in a colder climate, you might expect 15 percent less gas mileage.[21] If you combine factors for driving style and temperature, the actual gas mileage for a typical car rated at 26 mpg can drop to 14 mpg! Is the fuel economy sticker deceptive? With the continued revisions in methodology from the EPA and updated FTC guidelines (as recent as 2017), the new vehicle labels provide good, consistent data to the consumer. What is lacking is more information about the range and variance of the data and how it is compared between lab simulations and real-life experiences. To that end, the government encourages car owners to self-report fuel consumption on FuelEconomy.gov, and a simple online account registration makes this easier to do on a mobile phone at each gas pump stop.

We expect further improvement to this method will come in the near future with the connection of vehicles to the internet. Cars currently have onboard computers calculating fuel use and engine performance and soon that data might be routinely extracted and shared, creating a large data set of real-life vehicle fuel-efficient performance. As advocates for truth in data, the more data made available to consumers then the better informed they become, since one number like the *average* miles per gallon does not tell the whole story of the data.

Experts Recommend

Are there times when looking at data about a product can seem overwhelming? If you do not love data as much as we do, you may grow blurry-eyed when trying to make decisions about things you buy. Nuances in data collection and the proverbial fine print can be a bother to digest. Advertisers, of course, promote the benefits of their offerings and hope to move buyers as swiftly as possible towards a purchase. Numbers and analysis can hinder that process, so sometimes to move things along marketers introduce experts to provide recommendations. The value of the recommendation either as a testimonial or an advocate is to ease any concerns and perhaps to allow potential buyers to skip past the data and the hard thinking about the purchase and simply make a decision. This was exactly the tactic used in the 1940s and 1950s to sell cigarettes.

By the mid-1930s the cigarette industry had a serious problem. Cigarette consumption had started tremendous growth around 1920 and the industry was booming—a trend that would continue until the late 1960s. Simply put, cigarettes were popular and early advertisements showed the glamour and "cool-ness" of smoking by depicting healthy people and celebrities alike. Some were touting cigarettes for weight loss— "to keep a slender figure . . . reach for a Lucky [cigarette] instead of a sweet." The problem was that following the rapid growth of the number of people smoking there also began a rapid growth in the number of people dying from lung cancer. At the time, the data and science lagged from what we know today—smoking causes cancer—but there was a growing belief they were related. Customers were becoming skeptical and by the mid-1930s there was a decrease in cigarette consumption out of fear that smoking irritated the lungs and irritated lung tissue might be the cause of cancer.[22] Marketers' approach to this problem was a unique lesson in data deception.

The first concern marketers had to overcome was the irritability factor associated with smoking. The solution—data showing the new versions of their favorite brands were *smoother*. Innovations in filters and the apparent "toasting" of the tobacco in the curing process provided these tactile improvements to the smoke. Next was the inclusion of medical doctors' recommendations of particular brands—not only for their smooth, less irritable smoking traits but also supporting the health *benefits* of smoking. Clearly, cigarette marketers had not only taken on the dire concerns of their consumers but they also had pivoted the narrative. Cigarettes are not bad for you; they are actually healthy! An amazing feat that should probably go down in the marketing hall of fame of the biggest data dupe in advertising history—if such a thing exists. A 1937 Philip Morris advertisement in the *Saturday Evening Post* read "Men and women with irritation of the nose and throat due to smoking were instructed to change to Philip Morris cigarettes. Then, day after day, each doctor kept a record of each case. The final results, published in authoritative medical journals, proved that when smokers changed to Philip Morris, every case of irritation cleared completely or definitely improved."[23] There are so many deceptions in this ad it is overwhelming. Which is it, "cleared completely" or "definitely improved"? Are not all medical journals "authoritative"? Why were doctors keeping records? If this were a modern ad, and you were a data-duping sleuth, you might have already been asking the same types of questions, and the part about the doctor's role is a particular standout.

Some ads including doctors broadly claimed "more doctors" recommend a particular cigarette brand. To improve the authenticity of their claim the ads became very detailed. Lucky Strike cigarettes noted 20,679 physicians agreed theirs was less irritating, while RJ Reynolds promoted 113,597 doctors favored Camels. Some very specific numbers. How did cigarette companies win over so many physicians unequivocally? One method was how they conducted their surveys. In addition to advertising in medical journals, cigarette makers like Philip Morris would also attend the American Medical Association (AMA) conventions, set up booths, and hand out free cigarettes, and then of course ask doctors to fill out surveys.

By 1953, there was growing scientific evidence of the dangers of cigarette smoking and cancer.[24] Faced with data, the cigarette industry rallied together and published a statement known as the Frank Letter.[25] The editorial-styled advertising showed us another lesson in data duping. Confronted with data linking their products to cancer, the Frank Letter sought to raise distrust and doubt in the numbers (this is a common tactic even today, for example, there are data debates about climate change). Without harshly attacking the study, they wrote "we feel it is in the public interest to call attention to the fact that eminent . . . doctors and research scientists have publicly questioned the claimed significance of these experiments." Further committing to their concern for the health of their customers, they wrote, "We . . . always will cooperate closely with those whose task it is to safeguard the public health" and finally concluding they will create a Tobacco Industry Research Committee, which will fund additional research, presumably about the safety of their products. In reality, the committee continued to foster distrust through the creation of additional studies that conflicted with the notion that smoking causes cancer.

The tactic was to fight data with (bad) data and to a certain extent, they succeeded. Many smokers did not want to believe smoking was dangerous so having some data justification to continue smoking was welcomed by them. By 1954, doctors began distancing themselves from smoking, realizing the conflict between a dangerous product and advocating for an individual's well-being. Also, the AMA no longer accepted cigarette advertising in their journals. However, with the murkiness of data the advertisers created, it was not until the early 1970s that cigarette consumption began to decline spurred by the Public Health Smoking Act of 1969, which required standard warning labels on cigarette packaging.

Better Medicine

How much would you pay for a miracle cure? If it truly was a cure for serious disease, then the amount might be endless, hypothetically at least. In real-world examples, we are not often faced with such dilemmas—miracle cures are rare and money is limited—but it makes us wonder about the trade-offs in costs for improved methods of treatment when we look at how medicine is portrayed in marketing advertisements. How much better is better and do the numbers help us answer that question?

In the early 1980s, drug manufacturer Bristol-Myers Squibb was collaborating with Sanofi-Aventis, another drugmaker, to develop a better aspirin. Aspirin was a common household over-the-counter medication with a long history for treating various ailments from headaches and fever to preventing heart attacks. Ancient societies dating back before the Greeks used extracts of willow bark to create a crude form of aspirin, but the version we know of today was developed by German drugmaker Bayer in 1899. Since then, it remains one of the most studied medicines in the world with many applications.[26] For Bristol-Meyers, they admired aspirin's quality as an anticoagulant blood thinner that has long been used to treat patients following a stroke or heart attack. They eventually developed a drug called clopidogrel in 1982, and it was approved by the FDA in 1997 when it began Phase 1 trials and was marketed as Plavix. The drug was a commercial success and following trials in 2006 sales began to soar. By 2010 it had become the second best-selling drug in the world with sales of $9 billion (for context, world pharmaceutical sales in 2010 were $888 billion). Part of the success was due to its efficacy while part was also due to how it was marketed, and to understand more, we need to look closely at the numbers.

Plavix was, in effect, a substitute for aspirin for treating patients with vascular disease, specifically those who had suffered a heart attack, stroke, or peripheral arterial disease (PAD). PAD is a circulatory problem that reduces blood flow to your limbs. These three conditions share a common cause from the increased buildup of plaques of fatty material that inhibit blood flow. The purpose of Plavix was to treat patients and prevent the reoccurrence of these vascular events. The early data supporting Plavix was compelling, showing more than a notable reduction in events, and soon marketing materials were touting the achievement.[27] But there was a problem with these numbers.

The first came from a study comparing the use of aspirin to Plavix. After all, if Plavix was a better aspirin, then how much better was a relevant question. The CAPRIE study compared the efficacy of aspirin and Plavix for the occurrence of these three vascular events. When combining the three outcomes, the overall benefit to patients was a relative risk reduction of 8.75 percent. Relative risk reduction (RRR) is a commonly referenced metric in medical studies and measures the difference between two treatment groups, in this case, the difference of a vascular event between patients treated with aspirin or Plavix. The absolute numerical difference was a meager 0.51 percent since 5.83 percent of those taking aspirin experienced an event compared to 5.32 percent who were treated with Plavix (0.51 percent is 8.75 percent of 5.83 percent). The number might not seem very big, but if you were facing a fatal event like the possible outcomes of a stroke or heart attack, the advantage of Plavix is a meaningful number. However, this is where the numbers require a deeper look and where there is a flaw in using averages.

To arrive at the total risk reduction of 8.75 percent, the three event types need to be combined into a composite outcome. Separately the study showed the risk reduction was by far the greatest for PAD at 23.6 percent, followed by stroke at 7.26 percent, while heart attacks showed a *risk increase* of 3.78 percent. In other words, aspirin was better than Plavix for stroke prevention so the risk reduction was negative 3.78 percent. PAD is less likely to lead to a fatal event compared to stroke and heart attack. Including the benefits of PAD overstates the overall effect Plavix provides for the more fatal events of stroke and heart attacks, which if equally weighted yield a relative risk reduction of 3.49 percent.

Further, it is important to look at the confidence intervals for each of the vascular treatments. In statistics, the confidence interval is used when a subset of results is taken to make conclusions about larger populations. Since medical studies cannot measure *every* patient, the confidence interval is a guide indicating the *true* measure is within a high-to-low boundary and is stated to have "95 percent confidence" that the true number is within these bounds. When we look at the Plavix versus aspirin data there are relatively wide confidence intervals for heart attacks, which crosses what is termed the "midpoint of efficacy"— indicating sometimes patients did better or worse while taking Plavix. Again, using the *average*, heart attack patients were better treated with aspirin. The data for stroke follows a similar pattern—the confidence interval crosses the midpoint—although *on average* Plavix seems to be the better treatment.

The data from the CAPRIE study and others that followed provide a vast amount of information. So why did we dive so deep into the numbers? First, it is a lesson in applying the data-duping strategies from earlier chapters. For example, we see in this case the difference between the average and the range of outcomes as demonstrated by the confidence interval. The conclusions from looking only at the average may be misleading. The second is how this data story is retold in marketing.

Unlike other examples of data duping due to sloppy data curation, we want to be clear that data related to Plavix's efficacy is not a result of poor data collection or analysis. The data appear to result from well-designed academic studies. Like most data, it requires interpretation and the motivation of the people doing the interpretation sometimes requires scrutiny, and such is the case with the Plavix marketers once they had the data from comparison with aspirin.

What did Plavix marketing do with the data? They choose to *aggregate the averages* of all three events (stroke, heart attack, and PAD), and we can only speculate that with the inclusion of PAD, the numbers just looked better in their advertisements. After all who reads the fine print at the bottom of print ads, or reads those "disclaimers of doom" at the end of most medical TV commercials? As we noted this contributed greatly to sales. But what was the cost-benefit trade-off for customers? Aspirin is relatively inexpensive, around three cents per tablet. Researchers estimated that Plavix treatment to avoid one secondary stroke cost $33,000 compared to aspirin at $1,400.[28]

Further advertisement for Plavix wrote, "Plavix is the only *prescription medicine* of its kind that helps protect against both [stroke and heart attack]."[29] Note we added the italics to point out it was a factually true statement at the time, however, the data from the CAPRIE study also showed aspirin was also a protective medicine, just not one requiring a prescription.

In the end, these data tell us patients who are concerned about stroke (or likely reoccurrences of stroke) or PAD might find Plavix is a better treatment, while heart attack patients might consider aspirin as a treatment.

The good ending of this data story is since the introduction of Plavix and years of study compared to alternates such as aspirin and others for vascular disease, a common practice is to use multiple medications as treatment. Following the data can result in a more customized treatment plan—heart attack patients might favor something other than Plavix alone. The data used by Plavix marketers was deceptive, but the queries it raised with researchers and medical providers eventually contributed to better outcomes, and perhaps more transparency in advertising. As always it is best to talk with your doctor, ask specific data questions, and follow their advice.

COVID-19: A WORLD PANDEMIC

The pandemic resulting from the COVID-19 virus is a case study crossing many themes of this book. There was at times plenty of information (data) about the spread of the virus, its infection rate, death rate, and recoveries. Data about hospitalizations and capacities of beds and ventilators were also gradually available. Worldwide, governments had to make urgent decisions about stay-at-home orders, closing and then reopening services such as public transit and businesses. These decisions, of course, have implications for people's health and, as we learned, the survival of the most vulnerable people. Data and information spread quickly—some of it based on fact, others on opinion, and sometimes decerning the difference was not easy. The lines of fact and opinion are sometimes blurred and made it difficult for leaders to make decisions. We will explore a few challenges in both the math and the sensationalism that occurred with data and uncertainty during the pandemic.

A More Numerate World

In 1988, John Allen Paulos wrote the book *Innumeracy* with the intent to show people how a little application of statistics and probability could be applied to many everyday situations. He wanted to remove that cringe feeling many have when discussing math or casually reading numbers in the news. If there is one silver lining from the pandemic it is how many more people became more in tune with data and metrics.

As a nation, we were becoming more numbers aware as we tracked the COVID outbreak. Rarely did a day pass when people did not see a reference to the "numbers"—charts, tables, seven-day moving averages, long-term trending, and of course commentary followed. What did all these numbers mean and how were they going to impact our lives? After all, we were mostly stuck at home with little else to focus upon and soon even your uncle Dilbert was attempting a forecast from all those numbers. In the early stages of understanding a problem, like the spread of COVID, it can be challenging because there is limited and sometimes unreliable information.

When an Iceberg Is Really an Iceberg

If you were an ancient Viking sailing across the North Atlantic, you might have some experience bumping into icebergs. After a few encounters that experience would tell you most of an iceberg is beneath the water, with only a small portion showing above the surface. About 87 percent of an iceberg is unobservable.[30] Soon you would use this experience to change direction and avoid the large, previously unknown parts of the iceberg. Your knowledge and experience told you to rely on the small parts of the iceberg you could see to make decisions and avoid the parts you could not see.[31]

In the early days of the pandemic, there was limited information. And like an iceberg, observers of the data were attempting to see something larger than what they could observe. How is it spread and how quickly is it transmitted from person to person? Where is it infecting people and how deadly is the virus? How can the spread of the virus be slowed and prevented? Answers to these types of questions were important to medical providers and local governments. First reported in Wuhan, China, in December 2019 and soon thereafter in other parts of the world, the virus was yet to be labeled a pandemic given the relatively small number of known cases. At this point it was the "tip of the iceberg" in terms of data—limited reporting and the inconsistent diagnosis made the data difficult to be useful. For most observers, COVID was a curiosity for those living far from the outbreak. As the numbers increased in terms of reported cases and deaths, there was a growing need for data. John Hopkins University created the Coronavirus Resource Center and began providing the most frequently referenced information with daily updates of new cases and deaths from around the world. They gathered and consolidated information from hundreds of state and local sources. Yet, it was still problematic since the data were collected inconsistently and reported through multiple sources. The task remained—what could the small amount of data tell us about the larger spread of the virus?

The concept of data sampling is common in statistics and widely used in industry for tasks such as quality control in manufacturing. Sampling is when a small number of measurements are made from a larger group. In most cases, it is impossible to measure the entire population of something. In a large factory it is not practical to inspect and measure *every* product before it is shipped, so manufacturers rely on sampling—choosing a small number of products at random and measuring them. Applying statistics to the sample, conclusions can be made about the overall population, all the products being produced . . . within a certain level of statistical confidence. This concept was popularized by the Guinness Brewing Company back in the early 1900s when William S. Gosset developed a method of sampling that allowed for a small number of data observations called the Student's T-Test. Gosset's work allowed the factory to minimize the number of finished products removed for testing (and therefore unsellable) while making quality assessments of the larger number of their final product. Applying a similar method might tell us a lot about COVID using the small amount of data that was available at the beginning.

By March 24, 2020, there were 61,000 reported COVID cases in the United States and 1,040 deaths. That's a fatality rate of 1.7 percent. In little over a month, by the end of April the number of US cases would grow to more than 1 million with more than 65,000 reported deaths, a striking fatality rate. What was going on with the numbers? Was the virus getting more dangerous and, given these numbers, what is a rational response?

Of course, the fatality rate was not a symptom of a more deadly virus but rather an increase in reporting deaths attributed to COVID relative to the number of total cases, which was also shifting over time due to the availability of testing. Further, there is evidence the people who were tested were those who were seriously ill and had other comorbidity factors. Bias in the testing of more vulnerable patients overlooked the less serious nonfatal outcomes of others.

Nonfatal outcomes, although good news, contributed to misleading data about the virus. Many cases of COVID were not deadly, and the fatality rate was eventually known to be less than 2 percent, allowing for many cases, certainly before testing was available, to go undocumented.[32] For those who were tested and results documented, the data painted a picture of a more troubling situation. As a result, in the first few months, the US COVID fatality rate at times appeared, inaccurately, to be as high as 5.9 percent. Later, by August 2020 the rate was 3 percent and eventually trailed lower. A better view of the fatality rate would have been to separate the sample of test-positive patients between those with other comorbidities and those without other underlying risk factors. The Centers for Disease Control using this approach would later publish information showing nearly 80 percent of deaths were among people older than age 65.

Early in the pandemic, the inconsistencies in the data were mostly treated cautiously, however, some media reports did have a data-duping tone. For example, on April 2, 2020, Bronx News 12 reported "Officials: New Jersey's COVID-19 death rate spikes, totaling 537." NBCnews.com wrote on June 25, 2020, upon the discovery of 47 more cases in Texas, "Texas pauses reopening as hospitals inundated with 'explosion' of COVID-19 cases." The phrase "explosion" may have been a bit much given Texas had already experienced several days with similar numbers. Grimly, Texas would later face more days with daily death totals much higher.

Sorting through the early data proved difficult and some local governments felt bewildered, such as the report on June 26, 2020, "Nevada is experiencing a COVID-19 surge — and state officials still can't say why." "Can't say why" sums up the difficulty in making inferences with small data sets (relative to the larger population), especially when the underlying factors contributing to the spread of the virus and negative outcomes were somewhat unknown. Like the early Vikings sailing in the dangerous North Seas, they could navigate the weather and treacherous conditions and learned to avoid icebergs, whose hidden danger lay beneath the surface. Knowledge to interpret the unknown based only on what they could reasonably observe helped lead them to their best outcomes. A sample, observation and interpretation of data, can give insight into the larger population. With this though, it remains puzzling to us why the *Titanic* struck an iceberg anyway, and perhaps that is a more telling tale of the world's initial response to the pandemic.

Exponential Power

The math behind an epidemic is fascinating. Viruses do not just spread from one person to *exactly* one other person, they can spread from one to many, just like how a rumor is spread. One person tells another, that person tells a few others, and so on. Quickly it spreads. In this manner viruses spread to others at an exponential rate.

In epidemiology, simulation models are often used to understand the spread of a virus through a community. Models range from simple to complex. A simple model called S-I-R looks at three groups of people—those susceptible, infected, and recovered. The susceptible population is in general all people in the population. In a virus model, recovered individuals have gained immunity and are no longer part of the susceptible population. There are more complex variations of this model such as those that break out susceptible individuals between those who are exposed and those who are not, or further split infected between those requiring hospitalization or not. Further, components can be included to factor in the impacts of medical intervention treatments on the infected and recovered groups.

The initial COVID modeling was basic and suffered from fragmented information about the coronavirus, such as its incubation period and infection rates, and a biased measure of its fatality. The result in the media was, in hindsight, an overstatement of the threat and burden it would put on health systems. For example, on March 24, 2020, the *New York Times* reporting on the New York governor's daily briefing stated that New York City would "need up to 140,000 hospital beds to house virus patients . . . up to 40,000 intensive-care beds."[33] Based in part on these forecasts, the US Navy dispatched its hospital ship, USNS *Comfort*, to New York City, arriving in late March. The USNS *Comfort*

is routinely sent to disaster areas around the world and the virus outbreak was appearing to be a disaster. In addition, the US Army set up
an emergency hospital in a city convention center. A few weeks later,
however, the number of hospitalized patients was manageable and the
USNS *Comfort* left New York. There were similar stories in other parts
of the country. New York was not alone in its initial forecasts, so what
made it so difficult? In short, humans.

Forecasting the spread of a virus in an epidemic is a bit like forecasting the weather and the path of a hurricane. Many factors can influence
the outcome—the direction and severity of a hurricane can be influenced by other nearby weather systems in addition to air and water temperature. Hurricane models are extraordinarily complex and have had
the benefit of decades of refinement. The result? Even the best models
are faulty looking 10 days ahead. The best forecast can show likely
paths a few days in the future, which we see as the "spaghetti charts"
and forecast cone of the most probable path. Forecasting weather is easier than forecasting epidemics in part because the elements of weather
follow the laws of physics. Hot air rises and cold air falls at predictable
rates. However, viruses spread through the interaction of humans, and
that is where it becomes difficult. People do not always behave in predictable ways. Their individual social networks—size and interactions,
along with their efforts to stem the spread and their individual immune
response complicate models that otherwise assume a uniform distribution of humans and their viruses. Stated another way, the average infection from one person to a group of others cannot be assumed. Some
people may spread a virus to a small number, while others may spread
it to a larger group, becoming "super-spreaders." In our observation,
we did not see much discussion about the methodology of forecasts,
its range of accuracy (confidence intervals), or commentary about the
limited experience in modeling the spread of pandemic viruses across
large and varied populations.

Trends Cannot Go On Forever

Now that we have a model, even a basic one, what does it tell us? Can
the growth of COVID go on . . . forever? Well, no. As we learned in
the early chapter about the risk of assuming continuous trend growth on
the world's human population, we built an understanding that there are

always other factors other than the trend line itself. Those factors will constrain the trend from continuing forever. Human population growth's recent trend over the past 100 to 200 years was influenced more by improving education and health, factors that by themselves are not linear inputs that result in more humans. More education and improved medical treatments led to higher infant survival rates and longevity. It would also be a misunderstanding to expect more and more years of required education will result in more children being born. Similarly, with the growth of the number of COVID cases, factors such as changes in people's behavior would impact the infection rate. In other words, there are limits to growth, and much of it is driven by human behaviors. The expectation of a long trending line following laboratory-precise exponential growth was unrealistic as it was originally portrayed in the media. The more realistic and difficult task was incorporating the human factors into the trend lines to project the "peak" of the outbreak and when the trend line would flatten and eventually turn downward. Initially, as a trend line the number of cases seemed to be a straight line rocketing up forever, while time would show the reality was more like ups and downs of a roller coaster. Those first few months of data might have left the less experienced data duped about how the pandemic was unfolding.

Making Decisions in Uncertainty

The pandemic was an extreme example of making decisions in times of uncertainty and, in some regards, it may not be too different from how people make other decisions when faced with imperfect and lagging information. The implications of those decisions in hindsight were enormous as companies large and small had to decide how and if they would continue operating as a business. Governments at the national, state, and local levels were faced with similar decisions that were further compounded by the anxiety of their constituents. Policy decision could have real impacts on people's livelihood and possibly their lives.

On March 27, 2020, the *New York Times* headline read "Job Losses Soar; U.S. Virus Cases Top World." With it they displayed a profound graphic filling nearly the whole front page of the number of unemployment claims in the past week, which had increased 10 times more than the weekly average. From 345,000 unemployed to nearly 3.3 million!

As data professionals we admired the data visualization yet grew concerned about its implications. The change was more than the peak of 665,000 during the 2008–2009 *"Great* Recession." It was shocking, perhaps intentionally so. Worse, it was just the beginning. For each of the next five weeks, initial unemployment claims would be 3 million. Each week! The total number of unemployed would peak at more than 20 million by the end of May. Was this a data dupe? If so, what was missing? Within the article came more context, which gave it perspective. Most people just do not really know if 3 million unemployed workers was a big or small number. The comparison to 2008 and 2009 helped add perspective. There are 165 million people in the US labor force and eventually, nearly 15 percent would be unemployed. The better context was the unemployment rate, which was already historically low at 3.5 percent in February 2020. The increase of jobless claims increased the rate to 4.4 percent in March. What made these data worthy of the attention of the *New York Times* and its readers was the swift turn in unemployment. In other words, this was not a bunch of "malarkey." Although it was really the subsequent weeks of unemployment claims that had a meaningful impact and eventually drove the unemployment rate to 14.8 percent.

While unemployment rates increased other decisions were also influenced by the early trends of pandemic data and its interpretations. Businesses paused. All nonessential work paused. People paused. Factories once busy and in short supply of workers just a few months before were idle. Products were not produced. Consumer spending fell 30 percent.[34] Restaurants and the hospitality businesses decreased by 50 percent and furloughed employees. Schools and government offices closed. The world paused, largely based on the ambiguous data about how the virus spread and the harm that it might cause.

We are not being critical of the decisions that were made in early 2020. Many were in fact quite right given the limited information, if only inadequate on predicting the timeline in which the virus would eventually contribute to 95 million US cases and more than 1 million deaths.[35] We use this as an example of how important data are in making decisions and how significant those decisions based on limited information and forecast models can be to everyone. Early forecasts should be weighed with caution and the lessons of data duping we are advocating. Ask questions. Seek to know more about assumptions and understand probabilities and factors that could sway a model's prediction dramatically. Balance this with the cost of decisions. Wearing masks, for example, is a low-cost and simple remediation, even if a model is incorrect. In contrast, decisions to defer other life-limiting medical care may result in unintended consequences. For example, in April 2020 researchers noted cardiac diagnostic testing decreased by 64 percent compared to the prior year and suggest this contributed to an increase in cardiac-related deaths due to a lack of identifying and treating heart disease.[36]

The data and analytics of epidemics is intriguing and now we know also difficult. Our experience in forecasting the effects of a widespread virus and human behavior impacts were limited. Models showed, as expected, that social distancing was effective. Models also show the expected number of additional infections and deaths that could have occurred without social distancing and masks. In Paulos's 2001 revised introduction to *Innumeracy*, he noted in the period since his first edition some numeracy in US media had improved, and yet it showed opportunity to go further. He discussed the statistical anomalies of the 2000 US presidential election in Palm Beach, Florida (poor ballot design caused votes to be mistakenly cast for a third candidate—enough to change the outcome). He pointed to the media reporting of the Florida results being authoritatively explained by lawyers and journalists rather than statisticians. A few statistical techniques and a clear data visual might have been more useful for the media. Twenty years later with the media transcending from traditional media to social media and a greater motivation to provoke readers, there is little space for "statistical commentary." We have become accustomed to reading and seeing probabilities in weather predictions. Hurricane tracking during a storm has us all listening intently to data scientists and weather forecasters with advanced degrees explain the models and, as a result, help make life-saving decisions about evacuations and preparedness. It serves as a good example of using data effectively while being transparent about statistical uncertainties.

Our ideal for the future of media, might it be a disaster event, a pandemic, or a story about forecasting the fall harvest, is that data are more often presented with context rather than provocative headlines. More insights about the numbers and how they can impact our decisions—do we need an umbrella for our walk, or does our factory need to close this week? What are the probabilities of these events? Numbers by themselves can be deceptive. Data, as they say, is messy, fraught with problems in curation, and sometimes unexplainable ranges and outliers. Media and numbers, as we have discussed, can be a dangerous data-duping combination. Without the view of statistics in media, it is a bit like a medical doctor making a diagnosis using only one test. It requires more examination, input, experience, and of course science before a conclusion can be made. Fortunately, statistics has a long history of tried-and-true methods that can help interpret data, appropriately cast doubt where needed, suggest further research when prudent, and present statistically significant conclusions. Well at least within a 5 percent margin of error!

KEY POINTS

- There is a lot of "malarkey" in media throughout the news, marketing, and social media, making navigating data dupers challenging.
- News has a long history of provoking readership and the "free" press online allows anyone with an internet connection to create unfiltered and often unverified information.
- Marketing, although regulated to provide factful advertisements, still can have deceptive data.
- The COVID pandemic demonstrates both the rapid spread of information along with the challenges of interpreting data while working with uncertainty.

Everyday Decisions

WARRANTIES ARE LIKE CARNIVAL GAMES

Step right up! Knock down this stack of milk bottles with a ball and you win a prize. Only $2 to play! So goes the pitch from the carnival game attendant. Do you play?

If you have ever visited a carnival, you will notice everywhere you turn there are all sorts of games to play. Some require skills like throwing a ring over a bottle, while others are based more on chance such as the ball-and-cup-toss game. It is exciting watching others play. Occasionally there are cheers for winners and groans for the near misses. And indeed, every once in a while, you will see someone walking around with one of those overstuffed bears—a badge of honor for their carnival skills. But is it skill?

Our assumption at the carnival is skill is an important factor, but, like a Vegas casino, the skeptic in us knows the odds favor the "house" providing the game. What are the odds of winning some of those games and how much do you have to play to win the big prizes? It turns out the odds are low, just like any other game of chance, but you probably knew that anyway.[1] If you have ever been to a carnival, you have seen the milk jars, baseballs, basketballs, and rings are all a little different from the ones you may have at home. Even if you are skilled at say basketball, you won't be surprised when the carnival operator hands you an overinflated ball with some extra bounce to it. You are given an opportunity to win a prize, but you are not given much information about your chances, the effects of the modified equipment, and if $2 is a *fair* bet. Doubtful. Hopefully, most people play carnival games just for fun or to support a cause. Otherwise, it is a terrible way to invest a few dollars. Interestingly though, product warranties have a few things in common with carnival games.

Warranties and carnival games suffer from the same thing—information that allows the participant to make a good and informed decision. Warranties often are accompanied by numbers such as fees and deductibles but omit information needed to make a data-informed decision. Let's look at an example of extended car warranties.

Extended warranties come with different terms (years of coverage) beyond the standard factory warranty. This variety also makes them difficult to compare and a potential data dupe. We researched an extended warranty for a Honda Civic from several providers that ranged from $498 to $1,012 per year. The midpoint or median of the quotes was $671. Using this figure, a cost estimate of a five-year warranty is $3,350. It is important to note when you purchase a warranty you typically pay the full price upfront. Now, think of the warranty cost like a bet. You are betting $3,350 that you will have a repair cost in the next five years. Further, to make the bet good for you, you are also betting the discounted repair costs are more than $3,350 in total over the period. In other words, if the costs are less than what you paid for the warranty, then you have paid too much.

Consumer Reports performs an annual survey of more than 400,000 cars and trucks and reports annual average repair costs in year 5 and year 10.[2] We learn a few things from their data points. First, as expected, the repair costs for older vehicles are higher and lower for newer cars. No surprises there and of course more expensive cars often have higher repair bills. The year 5 typical cost is about $200, while the midpoint for year 10 is closer to $458. Conveniently Hondas are right in the middle of these costs and using an average for an estimate is reasonable. Further, we can also extrapolate the data for other years to get a rough estimate of repairs for any year after manufacture. The model estimated from the *Consumer Reports* information is as follows:

Annual Repair Cost = $51.64 × Years Since Manufacture − $57.92

Now, suppose you own a car that is three years old, the manufacturer's warranty is expired, and you purchase an extended warranty to cover five years from years four through eight. The cost of the warranty as we showed is $3,350. What are the expected average repair costs? Using the formula, the expected repair cost for year four is $148 ($51.64 × 4 – $57.92 = $148). Repeating this calculation for the remaining years and summing it up yields expected repair costs of $1,260. Judging by the averages it appears the warranty is overpriced compared to the expected repair costs.[3]

The data from *Consumer Reports* is missing an important element, which is probability. They show the *average* repair costs for each year by the manufacturer but do not provide us with the likelihood. Further, the average does not tell us a lot about the range of costs. If we knew repair costs might exceed the average by twice as much then we might reconsider the value of the warranty to cover this uncertainty. However, the information is complete enough for us to make a decision, but the absence of the number and frequency of repairs makes us unsure how often we will be heading to the mechanic, even though we know we will be paying for it ourselves. There is also a factor of inconvenience, which to some may have an economic value. After all, if we are running off to the mechanic once a month, even though the costs may be small, it may impact other parts of our lives and be disruptive to work and family and there is a cost to that too.

Let's explore another common example using consumer electronics. If you have ever bought a television, computer, or cell phone at a retail store, you very likely were offered some type of warranty. In our experience, these offers come at the last minute, usually while standing in the checkout lane with a line of shoppers waiting on you to make a decision. There you are, after spending a considerable amount of time researching your new purchase, reading reviews, shopping around at different stores to find the best price and they ask you something you likely know little about. This somewhat dubious practice of offering a warranty at the last possible moment, under time pressure, has not gone without notice. Researchers noted that as far back as 2003, both in the United Kingdom and in the United States, government agencies have been making recommendations about retail product warranty sales practices.[4] Notably, the prices for retail product warranties are not advertised and there is little information about their costs until the product sale is rung up. These practices make it difficult to separately shop for the pricing of the warranty since they are hidden until the end.

Government agencies such as the Federal Trade Commission (FTC) recognize critical decision-making information is absent from warranties. Their website is prescriptive in providing customers with clear information about warranty coverage and exclusions, all of which are required to be shown in writing.[5] However, there is no requirement to provide details of product reliability—how often it may need a repair or replacement, and the typical costs of those repairs. Without this type of information, how can we rationally decide the value of the warranty? In short, we cannot and we are about to be data duped. Depending on the salesperson you may hear anecdotal information about the cost of a recent repair. Keep in mind these estimates are likely to be extreme, those most memorable stories of the cost of repairing your product. Consider the source of that data, and remember this is not an objective random sample.

Back to the checkout line. There you are about to spend several hundred dollars on this product you have likely agonized over buying for some time, and you need to decide whether to buy a warranty, right now before that guy three people behind you who keeps checking his watch, shifting his weight in agitation, starts yelling at you to move it along. What do you do?

Fortunately, you are now armed with some of the data defense strategies mentioned earlier. You can make a few mental-math estimates: what is the cost of the product compared to the cost of the warranty? A warranty covering a full replacement that costs 20 percent of the product implies a likelihood of breaking 1 in 5 times. Based on your reviews from previous owners, can you determine if 1 in 5 is reasonable? If you can, you are on your way to a data-informed decision, and our bet is you will decline the warranty. Whew!

Declining to purchase the extended product warranty for consumer electronics is also supported by the data. Researchers in 2018 performed an analysis using the INFORMS Society of Marketing Science Durables Dataset reviewing 140,000 consumer transactions across 20,000 households. They focused their research on about 4,700 transactions involving televisions. They learned about 26 percent of consumers purchased an extended warranty and their average cost was 22 percent of the price of the product. Pairing this with data from *Consumer Reports*, they were able to show the failure rate, the rate at which the TV needed to be repaired or replaced. The failure rate also represents the probability of a claim. The average failure rate was only 6.7 percent. Let's do the math:

> Product Cost = $500
> Warranty = $110
> Probability of Replacement = 6.7 percent (100 percent loss)
> **The expected price of the warranty:**
> Product Cost × Probability of Replacement + estimated administrative costs

As we noted earlier estimating is an important data defense. It seems reasonable the administrative costs are between 10 and 50 percent, resulting in a Fermi-like estimate of about 20 percent. Calculating in this additional 20 percent results in a value of $40.20.

$$.067(500 + .2(500)) = \$40.20$$

If you knew ahead of time the probability your new product would fail and require a full replacement, then the most you should pay is about $40. So why did more than one in four people choose to buy the warranty? Part of the answer may be a result of the time-pressure sales tactics and the urgency to decide with a building line of agitated customers. Part of it is also our inability to make fair and accurate failure estimates, which the data show we tend to overestimate. Further, we may also be urged into making a decision to buy the warranty due to our loss-aversion bias. More on that later.

LOTTERY TICKETS

One day in 2003, Jerry Selbee, a retired former factory worker, bought $8,000 in lottery tickets for the Michigan Lottery. He allowed the machine to randomly pick the numbers and stood patiently at a convenience store as the lottery machine methodically printed out the 8,000 tickets. He won $15,700, a net gain of $7,700. He would later go on to *earn* hundreds of thousands of dollars in subsequent lotteries.

In 1992, Stepan Mandel and a group of his investors won more than $27 million in the Virginia State Lottery after purchasing nearly $4 million worth of tickets. For $1 each, they purchased every combination of the six-number lottery.

You may have heard other fantasy-like stories of lottery winners, including the $1.5 billion Mega Millions winner in 2018. However, the above two stories stand out because both Jerry Selbee and Stepan Mandel were not relying on luck, but rather on statistics and a couple of unusual defects in the games they were playing.

In the case of Mandel and the Virginia State Lottery, the rules were simple. Choose any six numbers from 1 to 40 in any order and if you match those drawn then you win the jackpot. If no one wins that particular drawing, then the jackpot is added to the next drawing. The defect in this game was as the jackpot grew and grew, it eventually reached a point where the winnings exceeded the cost of buying every combination of numbers.

For this type of lottery, the odds of choosing the correct number are 1 in 3,838,380. This is calculated as the total number of ways six numbers can be chosen divided by the number of ways of choosing them. It can seem confusing but consider when choosing a sequence of six numbers among the 40 lottery balls, the first is chosen from all 40, the next from the remaining 39, the next from 38, and so on. Without replacing the balls there are more than 2.7 billion combinations. (The math for six balls looks like this: $40 \times 39 \times 38 \times 37 \times 36 \times 35 = 2,763,633,600$.) Since the order of the numbers chosen is unimportant, that number needs to be divided by 720, the number of ways of picking six items ($720 = 6 \times 5 \times 4 \times 3 \times 1$). For the cost of $1 each and the backing of investors to bankroll the purchase, Mandel was certain to be holding a winning ticket on the day of the lottery.[6]

Jerry Selbee while playing the Michigan Lottery did not have to buy every combination of the pick-six game where players choose numbers from 1 to 49. He would have needed nearly $14 million to accomplish such a feat. What Selbee discovered in the Winfall game was each week that there was not a jackpot winner and the total jackpot was more than $5 million, then the prize money "rolled down" to other tiers where winners could earn money by correctly picking five, four, or three numbers. The odds of picking just five correct numbers out of six were better and it made the value of the tickets worth more than they cost. As long as you bought a large number of them, your winnings would be greater than the costs. How? The odds of choosing the correct six numbers are roughly 1 in 14 million, while the odds of choosing five are 1 in roughly 2 million. When there are 2 million possible tickets that could win $5 million, then each is worth $2.50 but only cost $1. Selbee figured this out for lower payout combinations as well and realized buying enough tickets, but not all, would eventually win more money than he was betting on the lottery.

Shortly after Mandel won the Virginia Lottery, they changed the rules of their games. Similarly, Michigan and other states with roll-down prizes changed those features. Most lottery games are similar to the popular multistate Mega Millions lottery, which requires players to correctly choose five numbers between 1 and 70 plus the Mega ball number between 1 and 25. The odds of successfully doing this is 1 in 302.6 million: $(70 \times 69 \times 68 \times 67 \times 66 \times 25) / (5 \times 4 \times 3 \times 2 \times 1)$.

The stories of Selbee and Mandel, who beat the system using numbers, are now legendary and seem to have resulted from poor design of games, and we might add, the innumeracy of the officials who oversaw them. Looking back, it should have been apparent there were flaws in the games, but perhaps they believed the sheer logistical difficulty of playing the odds to the full extent (i.e., buying 4 million tickets) was enough of a deterrent. Simple adjustments to the Virginia Lottery could have made the odds insurmountable to Mandel's strategy. Increasing the range of numbers that a player chooses from 40 to 50 would increase the number of combinations more than 10 times to 45,057,474. Alternately, adding the "mega ball" where the player effectively must independently choose another winning ball from 1 to 25 would increase the number of combinations to an astounding 95,959,500. Both adjustments would make Mandel's approach no longer economically viable since the jackpot winnings were never this large and the near impossibility of purchasing tens of millions of tickets.

The stories we see of people winning huge lottery prizes along with a few others like Selbee and Mandel can lead us to believe the chances of winning may *not* be too large for the prize money to be within our reach. In addition, some past winners profess they have proven strategies to increase your odds of winning, although we are suspicious of being data duped, so let's explore some of these ideas.

Winning Lottery Strategies or Data Duped?

Either browsing the internet or your favorite online bookstore will reveal there are many books and online guides to help you with winning strategies for the lottery. Some go as far as to claim they will guarantee to make you 10 times more likely to win the lottery.

The first common strategy is to pick your own numbers and use the same numbers for each drawing. This likely grows out of a distrust of automation and computers picking "good" numbers for you. Choosing the same numbers each time also does not affect your odds of winning. The only benefit might be that of regret if by chance your favorite numbers were drawn at the time you decided to pick new ones. Otherwise, a random number, either chosen by the computer or plucked from a random set of favorite numbers like the ages of your kids and birthdays, will have exactly the same odds of winning.

Another strategy is to *avoid* using birthdays since the range of numbers of course is from 1 to 31. There may be some benefit to this strategy *if you win* because other players have been picking favorite birthdays. So maybe by not picking birthdays, you are less likely to have to split the jackpot with others. Again, this does not help you win; it just likely gives you more if you do win.

Playing in groups with your family or coworkers is technically a way to improve your chances of winning, without spending more money on tickets. In this scenario, you would also have to consider the amount of your winnings is reduced because it is shared by the group. If you played with a group of 20 people and everyone paid for one ticket your odds of winning have improved by 19 more and your chances, in the example of Mega Millions, go from 1 in 302,575,350 to 20 in 302,575,350. Are those two numbers really different?

It is not much of an improvement but a little better. The probability changes from 0.0000000033049619 to 0.0000000660992378. According to our calculations for Mega Millions, the *average* jackpot, which is the sum of all the jackpots since 2002 ($24.897 billion) divided by the number of drawings (1,900) is slightly more than $13 million. Sharing it with your 20 coworkers brings it down to $655,000. Playing with a group can be fun, but it will not materially improve your odds of winning. A long shot, even improved 20 times, is still a long shot.

Another interesting strategy is to rely on information from past drawings. In fair games of chance such as a coin toss, roulette, or a lottery, the results of the past do not provide useful information to predict the future. This is such a common notion that it has been labeled the Gamblers Fallacy—the belief that past outcomes, like tossing a coin and getting heads six times in a row, can predict the next toss will be tails. Famously, at the Monte Carlo Casino on August 18, 1913, a roulette wheel, which has an even chance of landing on black or red, landed on red 26 times in a row. While on its way in this streak many bet and lost millions of francs certain in their determination red was coming next. Eventually it did, but too late for many who witnessed an improbable streak of 1 in 66 million.[7]

Along this theme the "experts" study past drawings and discover some numbers are more or less frequent in drawings and could be used as a successful strategy. The flaw in this logic is in a game of six numbers like Mega Millions, there are more than 2 million combinations of the first six numbers alone, and given the lottery has only existed since 2002 there have been less than 2,000 drawings. With a small fraction of the number of possible drawings (0.094 percent), the history of numbers tells us very little. This is similar to rolling a pair of dice three times and making a determination about the next roll based on the numbers so far.

One guide we found also suggested choosing six lottery numbers so their sum was between a range of about 110 to 195 since 70 percent of previous winning numbers were also in this range. The truth is, again previous numbers do not help with picking future numbers and further, *all* combinations of numbers, when six are chosen from 1 to 70, will have a 68 percent chance of being within these bounds. In other words, winning and losing numbers share this characteristic.

Got to Be in It to Win It

This is a favorite saying especially when there is a big jackpot and everyone it seems is rushing out to buy a ticket. We agree you have to own a ticket to win, but do you have to *buy* a ticket to win?

We found several stories of lottery winners who won on tickets that were gifted to them. In October 2019, Taylor Russey was working as a bartender in Missouri when one of the regular customers, who routinely buys tickets for the bar staff, handed her a ticket as part of her tip. It was a $50,000 winner. Jacqueline Carter of North Carolina received a "Big Spin" scratch-off ticket from her son as a gift to start the new year in 2020 and won $100,000. Dan Schuman got a ticket from his wife in 2019 for the Virginia Lottery New Year Millionaire Raffle that won him $1 million. And for no special occasion, Melissa Spagnola's father gave her a New Jersey Lottery "100X The Bucks" scratch-off ticket worth $2 million. In each of these stories, the winner did not buy any other tickets. So, do you need to buy a ticket to win and are the odds of winning much different if you do? Let's look at some estimates of the odds.

Each year Americans spend about $70 billion on lottery tickets. At an average ticket price of $2 that is about 35 billion tickets sold. The greeting card industry reports that 6.5 billion greeting cards are sold each year.[8] If we estimate that 3 percent of the tickets bought are given as gifts, it is still a huge number at a little more than 1 billion. One billion tickets spread across 6.5 billion cards still gives results in a 16 percent chance that the next card you receive could have a lottery ticket inside. How does this impact your odds of winning? Not much. If you bought a ticket, you would have a 100 percent chance at the same very long-shot odds of winning. In the case of Mega Millions with winning odds of 1 in 302.5 million, your chances are 3.304×10^{-8}. If you rely on obtaining a ticket as a gift in a card, your chances are 84 percent lower (1–16 percent) at 5.2879×10^{-9}. It is less than buying the ticket on your own, but purchasing a ticket does not substantially increase your odds of winning. It is still very, very rare to win the lottery.

Winning the lottery is such a rare event and perhaps that is the reason we see it reported in the news. Between 2003 and 2019 there were only 174 Mega Millions winners out of 1,768 total drawings. Less than 10 percent of the drawings had a successful winner. Further, Mega Millions has been adjusting the rules of their game, making it more difficult to win, by increasing the range of possible numbers and also adjusting the range of the Mega ball. The odds of winning the Mega Millions jackpot in 2002 were one in 135,145,920 when players had to choose five numbers between 1 and 52 and a Mega ball in the same range of 1 to 52. Notable changes in 2013 and later in 2017 decreased the odds of winning to 1 in 302,575,350 mostly by changing the range of the five numbers you choose to between 1 and 70. One consolidation partly fueled by an increase in ticket prices was the median jackpot rose from $67 million between 2003 and 2013 to $162 million thereafter. Along the way, there was also the largest jackpot in history of $1.5 billion in October 2018. (See figure 5.1.)

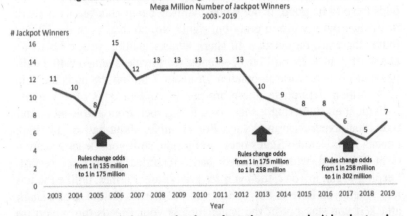

Figure 5.1. Changes in lottery rules throughout the years make it harder to win.

The Odds of Rare Events, like Winning a Lottery

With lottery winning so rare, it is difficult for us to give such a small number relevance. There are no comparable events in our lives with similar rarity. This often leads to others comparing winning to being struck by lightning—something we also perceive as extremely rare. The odds are in fact rare. Each year there are about 270 injuries from lightning and 27 of those result in death.[9] The odds of being struck by lightning in any given year are:

Estimated US population (2019)	330,000,000
Average lightning injuries including death	270
Odds of being struck by lightning in a year (1 in odds)	1,222,000

Compared to winning a lottery jackpot your odds of being struck by lightning may seem common. Your odds of being struck by lightning in any given year are 247 times *greater* than winning the lottery jackpot. The odds of being struck by lightning *twice* in a lifetime (1/15,300) × (1/15,300) are 1 in 234,090,000, which is still more common than winning the lottery. So, the next time you read in the news about a lottery winner you should also look for a story about someone being struck by lightning—twice. Odds are it happened!

Wait. We hear about people winning the lottery all the time so the odds have to be good, if not better than say being attacked by a shark or dying from a volcano eruption, right? No, not really. In the United States there are on average 40 shark attacks, putting your odds of an attack at 1 in 8.25 million. Volcanos worldwide average 540 fatalities and given a world population of about 7 billion the odds are 1 in 12.9 million. Of course, these are purely random odds of any given individual encountering a shark or a flying rock from a volcano. Your individual experience may vary. For example, about half of all shark attacks in the United States occur in Florida, so if your beach vacation is in Maryland, which has never had a recorded attack since record-keeping began in 1642, then you are pretty safe. Further, unlike the lottery where you have to be "in it to win it," there is little risk of a shark attack if you stay out of the water (unless your friends throw you in) and you just spend the day on the beach. But remember to wear your sunscreen! Laying on the beach can lead to sunburns and those with five or more sunburns are twice as likely to get skin cancer.[10] Your odds of melanoma skin cancer are quite high at 1 in 1,683, affecting 196,000 people in the United States each year.

For whatever reason, there are not a lot of statistics on other rare fatal events. The Centers for Disease Control and Prevention (CDC) and other government agencies track all sorts of causes of death, mostly in their fight against major diseases and infectious outbreaks, but they also keep track of some of the more unusual ones. Fortunately, it gives us a resource to look at the rare risks of death and ponder how much more likely they are than winning the lottery. Recall for reference the

Death From....	Total	Percentage	Random Odds
Staircase Fall	1600	4.85E-06	206,250
Bathtub Drowning	300	9.09E-07	1,100,000
Ladders	113	3.42E-07	2,920,354
Falling TVs	29	8.79E-08	11,379,310
Dog Attack	28	8.48E-08	11,785,714
Hippopotamus Attack	500	7.14E-08	14,000,000
Falling into Sand Hole on Beach	16	4.85E-08	20,625,000
Vending Machine Falling on you	2	6.61E-09	151,376,147

Figure 5.2.   The odds of picking the correct lottery numbers are 1 in 300,000,000.

odds of successfully picking the correct lottery numbers are about 1 in 300,000,000. (See figure 5.2.)

Lotteries as a form of public fundraising have a long history in the United States. Several of the Founding Fathers wrote of their need, including Thomas Jefferson, who wrote in 1826 "money is wanting for [public projects] a useful undertaking for which a direct tax would be disapproved. It is raised therefore by a lottery wherein the tax is laid on the willing only, that is to say on those who can risk the price of a ticket without injury, for the possibility of a high prize." Some lotteries were enacted but by the late 1800s they were under scrutiny. In 1868 Congress outlawed the use of mail for soliciting lottery players and in 1878 the Supreme Court wrote they had "a demoralizing influence upon the people." Legal lotteries did not return until 1963 when New Hampshire allowed a lottery to fund education. Gradually through the 2000s most states added lotteries, most touting the benefits to funding education. They have grown in popularity and public acceptance, with more than half of US households participating in the lottery.

There is a wide array of games from pick-a-number drawings such as Powerball and Mega Millions to instant-win scratch-off tickets. The industry generally does a good job publishing the odds of winning each type of game, however, much of this is lost in their marketing, which focuses more on the prize money and the glamour of being rich. With all those types of games to play it can be difficult to think about the odds of winning and it is easy to be data duped by those numbers. Reading the fine print helps you be more informed and if you are going to risk a few dollars, at least you will know the odds of winning are very small.

TRAVEL

Timing Airline Ticket Purchases

You are planning a big vacation for your family and you need to arrange the best flight itinerary and buy tickets. Your friends have convinced you there is a "secret in the numbers." You need to watch the trends and buy your tickets at just the right time. You have been checking prices like a day trader watches their stock prices ticking up and down. You have been making notes—smiling when the price drops a little, worrying when you see it tick up a notch or two. You set your alarm for the middle of the night because you have heard to "buy when no one else does." With the glow of your laptop as the only light in the room, you nervously search for your tickets and press BUY. Content you have made the best possible decision, you head back to bed. Have you found the best price and does it require the equivalent of a "night ops" to get it done?

The variation of an airline ticket price is based on the supply and demand of the available seats on a particular plane. Airlines try to anticipate our behavior—who will buy tickets early and who will wait until the last minute—and adjust prices accordingly. You may be a patient planner and buy tickets more often on the early side. Others who are either frivolous with money or last-minute procrastinators will buy tickets closer to the day of the flight. For whatever reason, they need a ticket and will pay nearly any price the airline requires. Often these are business travelers, and the airline industry has learned it can earn a good profit on these frequent customers who buy tickets at the last possible moment.

There is certainly a science to all of this, and airlines employ a complex set of analytics to optimize their revenues and keep their flights as full as possible. Our task as consumers is to try to understand the complexity of their numbers and use the data to our advantage to make the best decisions while avoiding being duped either by airline marketing to act with urgency ("sale prices end today") or the whim of the urban myths ("buy tickets on Tuesdays").

There is plenty of academic research on airline pricing models. Following the initial offering of an airline ticket, which is as much as 300 days in advance, fares increase when people buy tickets and decline if sales are slow.[11] The dynamic pricing remains relatively flat with slight variations in response to demand. Frequently they will dip further 100 to 21 days before departure and then rapidly increase. In some cases, the price can nearly double in the last 10 days before departure. The findings were consistent with an analysis by online booking agent cheapfare .com. Their analysis of 917 million airfares in more than 8,000 markets suggests the best time to book an airline ticket is 115 to 21 days in advance, although the price variation before then is only about $30. In the six days before departure, the single ticket price can increase by $220 or more.

Although that story about eating chicken soup for a cold may be true (and good for your soul), the one about what day of the week or time to purchase an airline ticket is not. In the study by cheapair.com they looked specifically for day-of-week trends and found that on any given day the price varied only about $2, or less than 0.6 percent. The mystery of how this myth got started is unknown but may have been twisted together with the true fact that the day of the week you travel matters. Tuesdays and Wednesdays are the least expensive, while Sundays are the most expensive. Likewise traveling over busy periods such as Thanksgiving or summer breaks will increase costs. It truly is supply and demand working efficiently.

Travel Insurance

Travel insurance is a good concept. In its basic form, the insurance pays you for any unreimbursed costs when your travel is delayed or canceled. If it were only that simple it may not be a topic for this book. There are two basic components to this decision: determining (1) if you need travel insurance and (2) how much you should reasonably pay.

Certainly, the cost of your trip is a known value—after all, you have to provide the costs of your airline tickets, hotels, and so on that you are wanting to be insured. So why is there a difference in the insurance rates? The answer is probability. Each company is making a slightly different assessment on the probability your trip will happen or not.

To make matters more difficult for you to assess and avoid being duped by the numbers, many travel insurance packages include add-on features. The add-ons have a value too, but individually they are tough to assign a number. Add-ons may include 24/7 emergency roadside assistance, a concierge to assist with rebooking flights, or assistance for lost luggage including payments to get replacements while traveling. Let's try to break down the basic cost of travel insurance and some of the probability of events you should consider.

The first thing to consider is why buy insurance? Are you protecting against something happening to *you* or something happening to *them*? The them, in this case, is the travel provider. The airline, a hotel resort, a tour company and the like. The first part will require you to assess your health and the health of those around you and think about situations causing you to cancel your trip. Planning to climb Kilimanjaro in six months? Then you need to consider what happens if you break a leg before your trip even begins. The second part is when something happens preventing the travel provider from delivering their service. This could include a tour operator going out of business or a weather event like a hurricane canceling a cruise. Reading the fine print for providers is also important here, since it may include information about how much of the travel cost is reimbursed and if it is refunded as a payment or a credit for future travel. Let's look at an example.

It is springtime and as you come out of your winter thaw you decide it is time to book a cruise someplace warm and sunny. You choose a 10-day luxury Caribbean cruise with a well-known cruise line for $5,000. To keep things simple, we will assume it is nonrefundable (in real life a decreasing portion of your fare is usually refundable 60 days before departure). For the first part, the "risk of you" resulting in a delay in your ability to travel might be the simplest. Let's assume you are otherwise healthy but your job occasionally requires you to work on a ladder. Your chances of a ladder fall may be low, perhaps estimated at 3 percent. Then the simple economic value of travel insurance would be 3 percent × 5,000 = $150.

Taking this further, let's also assume your planned cruise is in the middle of the hurricane season and a big storm would certainly cancel the trip. The Insurance Information Institute reports a typical hurricane season, which runs from about 150 days from June to November, could experience six hurricanes. What is a reasonable estimate that the 10-day cruise will occur during a hurricane and cancel the trip?

Again, we are using some of the estimation techniques we noted earlier. The point is to get a reasonable estimate to make a decision. A good way to think about the probability of a cruise disruption is to visualize a timeline. We will make some generalizations to keep this estimate simple. Assume that a 10-day cruise equates to 15 windows of time across the 150-day period, with any given window equal to the ratio of 10:150, or 6.6 percent. With a similar generalization, assume one hurricane disrupts travel for five days, resulting in 30 hurricane windows over the 150-day period, for a ratio of 1:30, or 3.3 percent. The cruise and hurricane are independent events and the probability of a cruise and hurricane occurring at the same time is combined percentages of those two events 6.6 percent × 3.3 percent = 0.2 percent. Extending this to the forecast of six hurricanes in a typical season increases the hurricane probability to 13 percent × 6 Hurricanes = 78 percent. Further, the probability of the cruise window overlapping a hurricane window is 6.6 percent × 78 percent = about 5 percent (or 0.0468). Again, we have made some generalizations about hurricanes not overlapping in time and that any hurricane would cancel all cruises, and we have arrived at a reasonable number of 5 percent. If there is a 5 percent chance of a $5,000 cruise being canceled, that value of cancellation insurance is $5,000 × 5 percent = $250. In total, the benefit of travel insurance to protect against you is $150, due to the risks of falling off ladders and so on and $250 for them due to the chances of a hurricane disrupting travel, a sum of $400. In reality, cruises are not canceled at the rates we have estimated here and this may be on the high side of the probability. Yet it serves as a good illustration of how to estimate how much insurance should cost given some rough estimates of probability. As a general rule travel issuance can cost 5 percent to 10 percent and aligns well with these estimates.[12]

IDENTITY THEFT PROTECTION

A recent radio commercial advertised their identity theft protection subscription with dire urgency and what initially sounds like a reasonable claim—"There was a victim of identity theft every three seconds in 2019." It sounded ominous and scary. Customer data breaches seem almost common and are highly publicized in the news. For this reason, the commercial gives us pause. "Every three seconds" does sound a bit terrifying, but what does it really mean?

- Does the reference to victim mean their data were stolen or that it was used in fraud? Often stolen data such as credit cards are made useless before criminals can do any damage.
- Every three seconds seems misleading. Identity theft is not like robbing a bank or purse snatching, which occurs one at a time. Identity theft usually occurs when a large batch of information is stolen. We hear about this in the news frequently. Was the statement "every three seconds" just sensationalized for the drama and to force some sort of urgency? In reality, it is probably more like 10,000 thefts occur in one moment and then none for an extended

period of time. So okay, *maybe* they meant on *average* there is one every three seconds. Although, the math still doesn't seem to add up.

- Checking the math. There are 31,536,000 seconds in a year (3.1536 e+7). If their claim is correct then at a rate of one identity theft each second that amounts to 10,512,000 cases.
- There are roughly 325 million people in the United States and about 255 million adults. Although we are not convinced about their claim of 10 million cases, the probability of any one adult being a victim of identity theft would be by their numbers about 4 percent (i.e., 10 million / 255 million). This seems small given the urgency of their message.

The math seems off and we suspect some misleading data duping in their claim. As a general rule, we are suspicious of any call to action that implores we need to "act now" before it's too late, before we are the next victim of our negligence. As we noted in the data defense chapter, this is an example where we need some known reference points.

First, looking at the small print at the bottom of the company's website will find the note that their claim is based on the Harris Poll. An online survey of 5,020 people was conducted and paid for by the company. Obviously with only 5,020 people in the survey they had to extrapolate the data to get to a national representation. Any time data are extrapolated there is a chance for more uncertainty, although it is possible the survey yielded 200 respondents (4 percent) who experienced identity theft.

The next fact check is also relatively easy since the Federal Trade Commission (FTC) maintains data and publishes annual reports on fraud and identity theft in their Consumer Sentinel Network Report. The report for 2019 shows a rather different picture.

In 2019, the Consumer Sentinel Network gathered information on 3.2 million reports of fraud and identity theft. Their information is gathered from various sources, including a reporting system that is exclusive to law enforcement. Identity theft claims can also be created by individuals on IdentityTheft.com. Of the 3.2 million reports, 650,572 were identity theft while the others were predominately fraud reports.[13] Indeed, very different from the claim in the radio commercial.

Identifying the disconnect between the FTC reports and this company's claim is impossible without knowing more about the methodology of the survey and how the results were extrapolated. Perhaps they only used the number of seconds in a working day (10 million) and maybe included fraud reports and identity theft as one category. Although if that was true, why would they not use the FTC report and chose instead to conduct their survey? Is it possible there is bias in the survey supporting their product? Again, without knowing how they arrived at the number it is difficult to determine. As consumers we hope to be rational, informed, and objective when making decisions, and when data are obscure, we suspect we are heading down the path of a data dupe. What might sound more reasonable? Using the FTC numbers (650,572) of identity theft the number is closer to one case every minute, and the probability any given adult would be an identity victim is relatively small at 0.26 percent.

Looking further into the numbers also piqued our interest about the amount someone should reasonably pay for identity theft protection. First let us say, being a victim of identity theft can be emotionally draining and stressful and there is no easy way to put an individual price on it, so we will not even try. Rather for this exercise, we will focus on the economics only.

According to the FTC Consumer Sentinel report among people reporting identity theft (650,572), 10 percent experienced a financial loss. The median financial loss is $700. Let's start with the expected loss calculation:

650,000 victims / 250 million US adults = a probability of 0.0026
× 10 percent (probability of a financial loss)
× $700

= $0.18

If you are 100 percent certain you will be a victim of identity theft then the expected financial loss is $700. How much you pay to either prevent that loss or be reimbursed should be the same amount, $700. However, you are not 100 percent certain to be a victim and the data suggests the number of victims is small at 0.26 percent. When the expected loss is 18 cents, then you should expect to pay something *slightly* more than that amount.[14] So what does identity theft protection cost? The basic plan in the radio commercial with loss reimbursement starts at $10/month or $120 per year. Based on the math and probability, this is looking overpriced . . . at least for its financial value.

PET INSURANCE

What is more popular than cat videos? Cat insurance. And dog insurance. The North American Pet Health Insurance Association (NAPHIA) reports pet insurance policies are growing at a rapid pace, increasing on average by 15.8 percent each year since 2015.[15] The number of policies has grown from 1.4 million in 2015 to more than 2.5 million. It is still only a fraction of the number of total pets in the United States, which is estimated to be 185 million. Most policies go to the dogs (83 percent)—perhaps there is some truth to "cats have nine lives" and their owners feel less of a need for insurance. The ASPCA estimates annual costs for a dog are $235 and for a cat at $160. These sound like reasonable numbers, so what is the value of pet insurance and how much should it cost?

Pet insurance is offered by a small number of companies, and like car insurance and warranties it suffers from the same type of flaws leading to data dupes—high-end estimates of unlikely expenses, missing frequency and probability of events, and overweighing emotional reasons to purchase. You may not have an emotional reason to insure your new TV, but when Fluffy winks those big eyes at you, it is easy to part ways with your money.

A typical pet insurance brochure will show you the expenses of various vet costs. Pet Plan (gopetplan.com) for example shows the cost of brain cancer at $14,000 and kidney failure at more than $7,000. Of course, what is missing is the probability. How often does a dog or cat have a treatable brain tumor or kidney disease?

Pet Insurance Review (petinsurancereview.com), which allows you to get pet insurance quotes from several providers all at once, goes a step further in their marketing. They provide the percentage of claims for each type of treatment. For example, 6 percent of claims for cats are eye infections and the cost is "up to" $6,000. A couple of things could be misinterpreted in these data. First, the percentages may be viewed incorrectly to imply these are the probabilities of such an event. They are not. Six percent *of claims* are for eye infections, not that 6 percent of cats will get an eye infection. Next is the wording of the costs "up to" a certain amount. These are not the average or typical (median) costs, but rather the *maximum* costs. Further, since most pet insurance policies have a deductible of $250 or more, the estimated costs, which are based on actual claims, are biased and will always be more than $250. What if the average cost for this treatment was less than the deductible ($250)? Knowing the cost can be up to $6,000 is not very useful in making your decision.

In our examination, we obtained 12 pet insurance quotes for our fictitious two-year-old small dog. Pricing varies based on age, size, and breed of dog. Quotes ranged from $7 to nearly $80 per month, although the $7 policy appears to be an outlier with a very high deductible and not comparable in features and benefits with the others. The median price was $46 or about $550 per year. We do not know the likelihood of an event so this is where your intuition and personal health assessment of your pet come into play.

Let's assume you expect at least one of these major events in the next three years. During that period, you will spend $1,650 (3 years × $550) on insurance premiums. Using our estimation strategy from our early chapter, we will estimate the cost of the one-time treatment by finding the geometric average of our high and low estimated costs. The pet insurance review information shows the highest cost is $6,500 (ligament injury or stomach issue), and we will estimate the low cost as $25. Really if you own a pet, you know anything lower than $25 is just not likely. The geometric average is $403. With a $250 deductible that you would pay first, the insurance would reimburse you $153. So far this is not looking like a good deal. You could have a few claims and still not receive more in benefits than the cost of your premium, but at least now you have a fair way to make an assessment. On the other hand, if you know your pooch will need more than their fair share of medical treatments, or has a penchant for chasing cars, then you might consider one of these policies. Just make sure they are not chasing the mailman. Although these plans cover when your pet gets hurt, they do not provide liability coverage when they hurt others.

WHAT'S IN YOUR HEAD IS PRICELESS

In the early 1970s, two researchers from Israel, who later would become lifelong friends, started to wonder why some people made irrational, that is, non-economical choices. Their hunch was that in different situations people would make different choices for the same problem. The "framing" of the problem was relevant and whether they were risking a gain or loss mattered. Their research delved into people's judgment and decision-making. Judgment is about how we make assessments of the probability and magnitude of outcomes, something people are particularly bad at doing and subject to our personal bias and the availability of information. Their work on decision-making is focused on how we make choices when there is uncertainty, and in particular how we favor avoiding risks, even improbable ones. By 1979, Kahneman and Tversky published their work titled *Prospect Theory: Analysis of Decision Under Risk* and the merging of economics and psychology into behavioral economics was born. As a foundation in this field, it led to works by Thaler and others and earned the Noble Prize in Economics in 2002.[16]

So, what does behavioral economics have to do with our everyday decisions on warranties, insurance, and lotteries? For one it helps explain why we do what we do even when it seems economically irrational. Not everyone plays the lottery, but some do. Not everyone purchases product warranties, but some do. And when people make those decisions, it is not always a result of misinformation. Sometimes it is about how choices are presented and the context of gaining or losing something.

If we offered people the choice between two outcomes:

A. 100 percent certain to lose $2
B. 0.01 percent (0.0001) chance to gain $20,000 or 99.99 percent to lose $4

Like the lottery, most people would choose B for the chance to win $20,000 even though economically the expected value for A and B are the same.

If you changed the lottery choices to bigger outcomes for choice B:

A. 100 percent certain to lose $2

B. 0.000001 percent (0.00000001) chance to gain $2,000,000 or 99.99999999 percent to lose $2

you very likely will get *many more* people to choose option B even though there is only a couple of cents difference between these choices. In this type of situation where there is a chance for a large gain, people tend to become risk-seeking. Kahneman later wrote in *Thinking Fast and Slow* that what people get from a lottery ticket is beyond its economic value, *"what people acquire with a ticket is more than a chance to win; it is the right to dream pleasantly of winning."* In other words, there is an emotional and entertainment value to playing the lottery. With small sums on the line, lotteries can be a cheap source of fun.

If playing the lottery can be chalked up to having fun and a dream, how do we explain people's affinity for warranties and insurance? In short, a nightmare.

Prospect theory introduced us to *loss aversion* and helps explain some of our risk avoidance behavior. When we are thinking about insurance for our pets or a warranty for our car, we are often confronted with thinking about our worst nightmares. On top of this, as we noted earlier, we are shown extreme costs for future events—my cat's kidney failure $6,000, a new transmission for my car $8,000. How could we afford such things? As the offer usually goes, "for a low monthly payment." As Kahneman explains, *"people buy more than protection against an unlikely disaster; they eliminate a worry and purchase peace of mind."* For the cost of insurance or a warranty, your nightmares can go away.

Prospect theory also showed us losses are more painful than gains. How much more? On the order of two times. That is the amount someone would have to gain to offset a potential loss. Offer someone a 50/50 coin toss to gain $100 or lose $100 and more people would not take the bet until the gain was $200 versus a loss of $100.

In this chapter, we have shown in many situations the data needed to make optimal decisions can be absent. Do you purchase the extended car warranty even though you have no data to know the likelihood of a major repair? To avoid being data duped requires some informed estimation to fill in the blanks for missing numbers and if you are lucky, a little research may help too.

Also important to our decision are the emotional connections. Whether it is the thrill of playing a game like the lottery or the calming release of worry from owning a warranty, each has its value in the decisions we make.

KEY POINTS

- Product warranties, including extended car warranties, lack needed information about event probabilities to make decisions on their costs. Marketing and sales practices increase the urgency to make purchase decisions. Research shows they are often overpriced for the benefits one receives.
- The probability of winning a lottery is extremely low, in the order of millions to one and you are more likely to be struck by lightning, be attacked by a shark, or find a four-leaf clover all in the same year than win the jackpot.
- Being misled by numbers and data duped in everyday decisions results from imperfect information, marketing, and myths. Our biases can also interfere with making rational decisions as we tend to overweigh small probabilities to avoid losses.
- Losses hurt twice as much as gains, so we tend to work to avoid losses more and overlook opportunities to optimize gains.
- Sometimes playing the lottery allows us to have fun and dream of winning big, and buying pet insurance makes us worry less. Once you know the numbers, at least you can make reasonable estimates, *and* if you can afford to dream big or buy peace of mind, then you should.

Making Big Decisions

> Wherever there is uncertainty there has got to be judgment, and
> wherever there is judgment there is an opportunity for human
> fallibility.
>
> —Donald Redelmeier, physician-researcher

BIG DECISIONS, BIG CONSEQUENCES?

"Big decision" is a relative term. Certainly, one measure of a big
decision is the magnitude of its consequence. Financially, depending
on one's wealth, one person's big decision might be inconsequential
to another. Other types of decisions are big decisions for everyone
regardless of wealth or status, for example, deciding on a strategy for
aggressive cancer treatment. Because big decisions are, well, big, it is
especially important to understand the risk associated with these types
of decisions and have a reasonable way to use data in your favor. The
bigger the consequence, the more sensitive people are to getting the
decision right, or at least not blowing it too badly.

A branch of analytics explicitly known as *decision theory* or *decision
analysis* tends to focus on approaches to making high-impact decisions.
For example, decision analysis is often used for capacity expansion
decisions in power generation and whether or not a pharmaceutical
company should use their patent to make a drug or sell it to another
manufacturer. Big decisions often require large initial investments and
have risk (uncertainty) about the outcomes. There are approaches for

reaching decisions that ignore uncertainty and others that incorporate probabilities to incorporate uncertainty.

Let's consider the following decision. One day you learn that you have inherited some land in Texas and have an offer of $150,000 to sell it. But, wait, you think there might be oil on the property and the company offering to buy it probably is thinking this too. After all, Texas has more than half of all the oil wells in the country, so why not?[1] A good way to approach this is to consider the possible outcomes. If there is oil it is a win, if not, it's a bust, so how can math and data help with this type of decision?

First, we know there are one of two decisions, drill on the land or sell it, and one of two outcomes, either there is oil or it is just a cattle pasture. Rather than blindly accepting the best offer, a good approach is to create a decision table. With two possible decisions and two possible outcomes, it is a two-by-two table—two rows for our decision and two columns for the outcomes.

The construct of the two-by-two table allows us to use some data, perhaps applying the estimation methods we mentioned earlier, to guide us. Let's start by applying some of those numbers. In our table, the decision row "sell" results in a guaranteed amount of $150,000 regardless of the outcome—oil or no oil. If we decide not to sell and instead the decision is "drill," then we might expect two possible outcomes. Drilling will cost an estimated $100,000 and if there is no oil it is a loss. However, if there is oil, perhaps the revenue we receive is $700,000. Is it worth giving up a sure thing of $150,000, or should you risk losing $100,000 to get $700,000? Maybe, and the answer is the next step of solving a big decision—probabilities.

The probability of finding oil in Texas can be as certain as the Permian Basin and Cotton Valley or unlikely as the oil-barren El Paso region. Researching that data would be helpful; however, to keep this example simple, let's estimate there is a 50 percent probability of finding oil. This allows us to calculate an important number—the *expected* payoff. The expected payoff applies the probability to each of the possible outcomes. It works like this: for each decision row ("sell" or "drill") multiply the outcome ("no oil" or "oil") by 50 percent and add them across the rows. Therefore, the decision row for "drill" results in an expected value of $300,000 (50 percent of –$100,000 + 50 percent of $700,000), while the "sell" decision expected value is a not-surprising value of $150,000. With the added benefit of probability and estimated

	Outcomes		
Choose	**No Oil 50%**	**Oil 50%**	**Expected Payoff**
Drill	($100,000)	$700,000	$300,000
Sell	$150,000	$150,000	$150,000

Figure 6.1 The decision table and expected value when there is a 50 percent probability of finding oil.

outcomes along with the construct of a decision table, it makes it easier to know when to say yes. In this example, the odds are in favor of drilling rather than selling at the offered price.

A great feature of decision tables is they can also be solved in reverse. For example, given the estimated outcome of finding oil and the cost of drilling, what would be a fair price to instead sell? $300,000 is clearly the minimum. Taking this further, how likely does it have to be that there is oil to switch from "drill" to "sell" based on the expected payoff? The short answer is when the probability of finding oil is less than 31.25 percent it is best to sell.

Applying probabilities to big decisions is very useful, however, those probabilities are not guaranteed. Probabilities are simply an attempt to estimate uncertainty, but uncertainty still exists.

To illustrate how we can better deal with uncertainty, let's move to a more common decision most people face, financial planning for retirement. (See figure 6.1.)

RETIREMENT PLANNING

Retirement is, of course, a big decision. When do you start saving and how much should you put away each month? How much do you need when you retire and how should it be invested—stocks, bonds, cash, real estate, those gold coins your uncle Dilbert is always talking about? The task is either economically difficult or so daunting that many just avoid it all together, leading to many saving too little too late for their golden years.[2] In short, retirement planning can be complicated and sometimes even emotional and for these reasons, much of retirement planning is beyond what we can cover here in these few pages. What we can explore is how many of us can be data duped by the math, and more notably the assumptions.

Beware of the Shark Fin

Retirement planning can be broken into two segments, and if you have tried one of the many online retirement calculators you know exactly what this means. The first is the period when you are saving money and your balances are growing. The second is after retirement when balances are declining. A typical calculation of balance growth and decline when graphed looks like a "shark fin" (see figure 6.2) cutting through the water, showing sweeping balance increases during the savings years and gradual decline through retirement. The graphs are beautifully elegant and simple, even reassuring, except they are completely wrong.

What makes many of these graphs wrong is how they portray some sort of certainty without considering how the actual outcome can vary. Take the first part of growing your savings over time and that upward slope of balance accumulation. If you have $10,000 how would that grow over 35 years? Looking at the real values of the last 35 years, the S&P 500 average annual growth was 9.9 percent from 1985 to 2020. We can input this into an online calculator or do the math showing

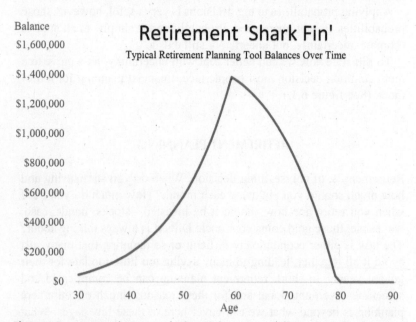

Figure 6.2.   Many retirement calculators show an unrealistic view of retirement balance growth and decline.

it could grow very nicely to $272,220.[3] We did this with countless online calculators and all had similar results. Each year it *consistently and evenly* grows upwards. What they fail to show is the reality that although the *average* annual growth is 9.9 percent it comes with many years of ups and downs and large variations from that average. In the period 1985 to 2020, there were 9 years of negative returns, the poorest was 2008 losing 38 percent, and there were 13 years of more than 15 percent returns including 34 percent in 1995. What this means is a simple projection of 9.9 percent growth may not easily get your balances to your goal. Considering the reality, the calculated growth of $10,000 invested in the S&P 500 in 1985 is $177,777.

How is it possible to be data duped and end up this far off our targeted balance of $272,220? It is a notable difference and a result of assuming all of your money is there working for you during the entire period. Simple online calculators (and even some financial advisors) overlook the compounding impact of a few mistimed down years in the market. For example, when the stock market had consecutive negative years beginning in 2000 through 2002, it took more than *seven years* for people's investments to return to their previous 1999 levels. That may be unusual, but a down year here and there can reduce the result, but happily, there are often more up years than down.

Now that we know the consequences of picking an average rate of return, what questions should we be asking of retirement planners, and is there a better way to get a realistic view of your retirement future? The answer is not to just choose assumptions about market returns from the past 5, 10, or 50 years but to ask about the variance of returns and the *range of possible outcomes*. History may not perfectly repeat itself but history does offer some statistical clues to help you plan the future. Varying returns from one year to the next should be expected, just like variances in spending after retirement should also be part of the plan—you might on a whim take that unplanned trip to Paris after all. Assumptions like market returns and anticipated expenses are very useful, but they must also come with a measure of their uncertainty. Failing to account for the uncertainty is how the numbers can deceive and how a retirement plan either using an online calculator or working with a retirement advisor can set unrealistic expectations. The solution is to ask questions about the *probability* of successfully meeting your retirement goals and under what scenarios can it be good or bad. These scenarios can be created with a few spreadsheet tools and a common method known as Monte Carlo simulations. More on the mechanics of that technique in a later chapter.

It's Your Retirement, Flipping the Question

How much do you need to save to retire is a very common question. Maybe we should flip it and ask how much do you need to spend in retirement. Maybe you would like to retire with enough money to fund 20 years of retirement. You also need to know how much money you should have on hand at retirement, the amount you expect to spend each year, and the rate of return on your investments and savings during your retirement years. For example, if you retire with $400,000 and can earn 14 percent you will be able to spend about $64,000 per year. Although a 14 percent return might be a bit unrealistic given the longer-term *average* of the S&P 500 is 10.2 percent. And chances are you will not be 100 percent invested in the stock market, further reducing the expected returns to something less than that average. So, be cautious about using the long-term stock average as a reference point since it may be too optimistic.

If you are conservative, you might think about what is the lowest return on investment you could experience over the 20 years. Perhaps you are willing to say 1 percent. What will that get you in terms of yearly spending during retirement? Using the time value of money methods mentioned earlier, it is about $22,177 per year. Ouch! By the way, this is ignoring inflation, which can further diminish your buying power.

As we are flipping this question, let's also consider the rate of return percentage needed to get to a particular desired spending amount. If your goal is to spend $64,000 per year for 20 years, and you only earn 1 percent per year, how much is needed at the time of retirement? The answer is roughly $1.2 million. It is time to start saving more for retirement!

It would be great if we could get a guaranteed 14 percent. But alas, life is not so simple or grand. The thing we are ignoring in our analysis thus far is that there will be *variation* in returns each year over these 20 years similar to what we showed during the savings period. The variation, the range of possible annual return percentages, is how retirement planning can be data duped. How do we solve for a multitude of possible outcomes and get to a realistic outcome? One method is to create a simulation.

Simulations can be used on a computer to measure a lot of "what-ifs," such as what if the return percentages were 14 percent every year, or what if the percentages were 5 percent some years and 20 percent others and still average 14 percent. How would all of these "what-ifs" impact the amount of retirement savings in the future? To give a sense of what this means, a starting savings amount of $400,000 given the right series of "what-ifs" in the future could become as much as $7 million. Although this is unlikely since it would take a rare series of market return percentages to reach that amount. The point of the simulations is to use data to look at many possible outcomes and conclude the most realistic are scenarios that tend to appear most often in the simulation. Stated differently, simulation allows you to estimate the probability of meeting your goal. For example, how likely is it that my retirement money will last at least 20 years? In our simulation, the money may last at least 3 years with a probability of 100 percent, 10 or more years with a 65 percent probability, and a 50 percent chance that our fund will last at least 13 years. Not every simulation works out this way, but the framework allows us to make a better informed, less duped decision based on how much risk we are willing to accept.

Certainly Uncertain

When you plan your retirement, many factors come with uncertainty. The longer the timeline between your plan and retirement further complicates things. For many of us planning one year or maybe five years into the future is something we can feel confident about, but with retirement we are often planning for 10- or 20-plus years down the road. A lot can happen in that amount of time and without the experience of our future selves to guide us, we need to look at the more uncertain variables in our retirement planning equation.

So, what really matters when planning for retirement? What are the real drivers influencing our ability to achieve our retirement goals? Based on our simulation, not surprisingly, the most influential factor is the age when you retire—when you delay retirement there is a double impact because you continue to contribute money and accrue interest and you delay withdrawing the money.

Most conversations about your retirement will require you to make assessments about all of these uncertain factors. Either in person with an advisor or using an online calculator, the goal often is to get to one number by providing specific inputs. What age will you retire? How much will you save each year? How much will you spend? And even, what rate of return will your investments earn? Creating a plan with such certainty—both the inputs and the calculated outputs—can lead your plans astray. The numbers are not so simple. They vary and, as a result, need to be treated differently to avoid a data dupe. The best approach is to know a range of possible outcomes and their likelihood. Armed now with the knowledge of tools such simulations of your retirement plans should both have a number (for example, how much you will save by retirement age) and a probability (the chance of realistically meeting that goal).

INVESTMENT DECISIONS—SKILL OR LUCK

Retirement investments are an important part of planning for the future but not the only "big decision" when it comes to money. Your non-retirement investments are important too. They might be a supplement to your retirement plan or the means that you use to save for other items such as saving for a home, a vacation, or planning a wedding. Similar to retirement planning you need to have a realistic expectation of how your money can grow. Through our research, we kept coming back to the deceptive use of numbers in investments known as survivor bias. We will review several types of data bias later, but let's look at how this particular bias can influence investment decisions.

Survivor bias occurs when we focus only on something that has made it through some process. One of the earliest and most well-known examples of survivorship bias occurred during World War II. The Allied forces decided to conduct a study to help protect their fighter pilots and planes. As the story goes, given the way pilots fly and the enemy fires, the idea was to find "hot spots" on the planes most likely to be hit by enemy fire. Maybe the hot spots could be reinforced with lightweight material that would better protect the planes and their pilots.

As planes returned from battle, the data on where bullet holes existed was recorded and overlayed. Statistician Abraham Wald noticed that most of the bullet holes were concentrated on the fuselage and very few were around the motors of the aircraft. In an "a-ha" moment, Wald realized that these data were from the planes that had *returned* from battle! They only had data on the planes that *survived* the battle. If anything, the hot spots from these data indicate areas that do not cause the loss of a plane.

So, what does survival analysis have to do with investing and being data duped? Again, the mathematician John Allen Paulos provides an interesting example in his book *Innumeracy*:

> Some would-be advisor puts a logo on some fancy stationery and sends out 32,000 letters to potential investors in a stock letter. The letters tell of his company's elaborate computer model, his financial expertise, and inside contacts. In 16,000 of these letters, he predicts the index will rise, and in the other 16,000 he predicts a decline. No matter whether the index rises or falls, a follow-up letter is sent, but only to the 16,000 people who initially received the correct "prediction." To 8,000 of them, a rise is predicted for the next week; to the other 8,000, a decline. Whatever happens now, 8,000 people will have received two correct predictions. Again, to those 8,000 people only, letters are sent concerning the index's performance the following week: 4,000 predicting a rise; 4,000 a decline. Whatever the outcome, 4,000 people have now received three straight correct predictions.
>
> This is iterated a few more times until 500 people have received six straight correct "predictions." These 500 people are now reminded of this and told that to continue to receive this valuable information for the seventh week they must each contribute $500. If they all pay, that's $250,000 for our advisor. If this is done knowingly and with intent to defraud, this is an illegal con game. Yet it's considered acceptable if it's done unknowingly by earnest but ignorant publishers of stock newsletters, or by practitioners of quack medicine, or by television evangelists. There's always enough random success to justify almost anything to someone who wants to believe.[4]

Now, it is not clear if this stock-picking scheme has ever actually been done. But it is clear that half of each iteration would have the correct pick and the lucky 500 that survive the previous rounds make the scammer look like an investment guru.

While Paulos's example may have been hypothetical, consider this real example of data duping among mutual fund investments. Mutual funds are great investments, especially for new investors who need to diversify their investment risk—one share of the fund gets them a small piece of every stock or bond in the fund. They provide professional investment managers and ease of daily investing or withdrawal. However, the US mutual fund industry, which has $21.3 trillion in assets, is also very competitively marketed to consumers.[5] Mutual funds are often ranked on their investment return performance, and survival bias plays a role. For example, long-term performance over 15 years can be accomplished only when a mutual fund has lasted that long, which seems obvious. What is less obvious is that mutual funds that do not perform well never make it to year 15.

Among a recent ranking of the best mutual funds of all time, all of the top funds existed since 1995 or earlier.[6] The top fund, Fidelity's Select Software & IT Services, which has an average annual return of 16.05 percent, started in 1985 and is invested heavily (25 percent) in Microsoft. Yes, that Microsoft, which has grown by more than 272,000 percent in value since 1985, or an annual rate of 24.6 percent.[7] This makes us wonder if the good long-term performance of mutual funds is due to skill or luck of an early bet on a few extremely well-performing stock investments. That is how survival bias works.

The data dupe is how fund rankings focus our attention on those that remain and does not give a full picture of all mutual funds past and present. The implied conclusion is "there is something great about these funds and we should invest." However, a 10-year study from 1992 to 2001 of mutual fund exits (closed or merged into other funds) showed among 7,500 funds that 27 percent no longer existed.[8] Twenty-seven percent! Of those, 30 percent closed in just two years, 2000 and 2001. By now, you are likely catching on that those were two terrible years in the stock market, in fact, the worst of the 10-year period, and as the industry has shown us, also the precise time poorly performing funds are merged with others performing better. This is not a coincidence. Another more recent study through 2020 showed typically 5 percent of mutual funds are merged each year with extraordinary fund exits during down markets in both the 2000–2001 and 2008–2009 periods, the latter striking a high point of 11.5 percent of funds exited or merged.[9]

Mutual fund companies are vying for the attention of individual investors through their marketing efforts. In part, this is because *most* mutual funds underperform the overall market. However, the practice of merging poor-performing funds and touting the surviving high performers provides unrealistic data. It is the equivalent of making investment bets on a few promising companies and, in the end, only keeping and reporting your successes. Although regulators such as the Securities and Exchange Commission (SEC) require mutual fund marketing disclaimers that "past performance does not guarantee future results," the advertising still touts recent performance as a badge of success, and in some ways, it is a data dupe.[10]

CHOOSING COLLEGE

Investments are not the only place where survivor bias can impact our decisions. Consider another big life decision about college—should people invest in a college degree and which schools are best? Your uncle Dilbert says "you don't need to graduate from college to be successful. Bill Gates didn't graduate from college and neither did Mark Zuckerberg. They are both worth billions!" Uncle Dilbert's statement is of course technically correct. You could come back with "yes, but Jeff Bezos is worth more than those guys and he graduated from college!"[11] But do you want to base your decision on whether to go to college on these outliers who *survived* to be billionaires? Rather than rely on survivors (okay, thrivers!—these three are pretty smart), maybe you need more data. Maybe the right question to ask is "how is income related to the educational level achieved?"

We obtained the median weekly earnings for different levels of education from the US Bureau of Labor Statistics, based on the 2020 Current Population Study, a survey of 60,000 US households.[12] The median weekly earnings for a high school graduate are $781, whereas a college graduate with a bachelor's degree has median weekly earnings of $1,305, a 67 percent increase. We note that Mr. Gates and Mr. Zuckerberg are actually both in the category "Some college, no degree" with median weekly earnings of $877. We're not sure if their households were included in the survey, but it doesn't much matter since the median is reported rather than the mean or average. Maybe college is not for everyone, but do not be duped into thinking it is likely you can skip college and easily become a billionaire.

TO BUY OR RENT?

As of February 2020, 64.4 percent of people in the United States owned their own homes. Millions of people every year face the decision "is it time to buy a home or continue to rent?" In 2020, there were more renters than at any time in the previous 50 years.[13] What are the factors that people rely on to decide to buy versus rent? Some factors are psychological—for example, a sense of ownership and security, and independence, but we will focus on the financial part of this decision.

Fundamentally the financial calculations are straightforward, the difference between the cost of owning and the cost of renting each year is either positive or negative. If the cost of owning the home is less than the cost of renting, and depending on how many years you plan to stay in the home, then it makes sense to buy. You have to factor in how long you plan to own the home so that you can determine if you will make up any of the upfront fixed costs of buying the home by the difference in cost each year. Of course, inflation matters and so the timing and magnitude of the annual differences matter.

What questions should we be asking? If I buy, how long do I plan to live in the house, and how does that impact the benefit of buying versus renting? So, the real risk in this decision might be how likely it is that you are going to sell in the first few years of owning the home.

FINDING A SOUL MATE

Finding someone to commit to for the rest of your life probably deserves to be in the chapter on big decisions. Can you take a data-driven approach to this problem? Maybe. Perhaps you have heard of the 37 percent rule. The 37 percent rule is based on the mathematical solution to the classic "secretary problem."

The secretary problem is described as trying to hire the best candidate for a job among a list of applicants without interviewing all of them. The challenge is by the time all are interviewed the best people may no longer be available to hire since in the meantime they found other jobs. How many people need to be interviewed before you know who is best for the job?

The answer requires a few assumptions. First, there is more than one acceptable candidate. Next, the candidates are selected randomly for interviews. Setting it up this way, after a certain number of interviews an assessment can be made about the whole group, including those remaining to be interviewed. The first group interviewed tells a lot about the next group of candidates. It can be shown mathematically that rejecting the first 37 percent of the total candidate list and then hiring the next person who is better than all of the previous ones maximizes the probability of finding the best candidate from the list. In other words, it is the best strategy to find the best person without interviewing forever.

Applying this method to finding your best soul mate, you might decide to date 50 people, not at once (!), but over time and in no particular order or preference, other than they meet your minimum wants in a person. To maximize the probability of getting the best soul mate from the 50, you would reject (stop dating) the first $.37(50) = 18.5$ (call it 19 people). Then, keep dating and stop when you find the person who is better than all 19 of those previously rejected. How simple is that? Well, it does assume a lot of things, including that you have a well-defined unbiased method to evaluate your candidates. If you like structured ways of making decisions, this might work for you too. This method can also apply to other decisions where there are many choices and limits on the practicality of evaluating every one, such as finding that ideal apartment, the best used car, or the perfect wedding dress. However, remember this solution is about optimizing time and maximizing benefits and it is solved just like how you would expect mathematicians to do so—with numbers! But if you believe in "the one, and only" when finding a partner, well it just might not work. There is a big difference between theory and practice. Let's investigate something maybe a little more practical and data-driven: internet dating.

Online dating services are very popular. Should you find your soul mate through the traditional approach of meeting people, dating through social networks, or using an online dating service? If we can measure success in finding a good soul mate by how many couples do *not* get divorced or by the expected length of a marriage, maybe we can see some differences. In fact, some studies have investigated this very question.

About one-third of marriages in the United States now begin online. One study over seven years found that 5.9 percent of couples who got together online broke up while 7.6 percent of those who met offline broke up. Of the more than 19,000-plus couples in the study who got married, only 7 percent separated or divorced in the seven years[14] (other studies have shown that about 20 percent of all marriages end in the first five years).

Making big decisions can be difficult and stressful. Taking a data-driven approach can often make us more comfortable assessing risk, which at the end of the day is what we need to do to make a sound decision. Our need for resolving these risks might lead us to seek simple solutions without question. However, with the uncertainty that often comes with big decisions, it is more important to ask the right questions. Question the data and the assumptions used in assessing the situation and make a sound decision based on your tolerance for risk and your understanding of probability. Doing this will reduce your chances of being data duped when it matters the most.

KEY POINTS

- Many of our big decisions are influenced by probability and often we humans are challenged to measure likely outcomes with certainty.
- When there is uncertainty related to our big decisions, the data can be deceptive. Working through multiple likely outcomes and scenarios using techniques like simulation can provide a better, more realistic projection of the range of possibilities. It is better to make decisions using this range than an absolute number since a range more realistically captures risk.
- When planning retirement decisions, ask questions about the *probability* of successfully meeting your goals and under what scenarios can it be good or bad. If you are given only one number for your retirement goal, ask again.
- Big decisions where many choices are impractical to individually evaluate can also benefit from a data approach, including selecting the best place to live or choosing a life partner.

7

Analytics at Work

How to Be a Better Decision Maker, Look Smart, and Impress Your Boss

DECEPTION

In 2003, Elizabeth Holmes had a brilliant idea. She recently had left Stanford University and was raising money for her new company. Her plan was a never-before-attempted approach to health care testing that would discover, and presumable preemptively treat, hundreds of diseases from a small single ampule of blood taken from a finger stick. The small amount of blood would be tested in a revolutionary desktop lab—a lab in a box—that would provide near-instant results. The process was simple. Collect the blood sample from a patient's finger, insert it into the box, and magically the results would appear in a few minutes. Did you have cancer or hepatitis? Are your kidneys functioning properly? In total there were 240 possible tests. The lab boxes were name Edison and styled to resemble the NeXT computers, an innovation in themselves that Apple founder Steve Jobs developed in the late 1980s. They looked magnificent. Inside was a complete automated lab—a fusion of robotics and cutting-edge laboratory technology. Their compactness and efficiency in which they could do testing, even without an order from a medical doctor, all drove down the costs and the convenience to customers up.

Now with this innovation, people who might not otherwise know they needed medical treatment could get it sooner, perhaps even before they showed signs of the illness. Theranos and their Edison lab machines deployed at local neighborhood pharmacies would, as Holmes often described it, "change the world." Who would not want that? Some may say they absolutely needed it. Some states changed legislation to clear the way for them to be distributed, sidestepping long-standing medical practices requiring doctors to consult with patients prior to ordering tests. After all, this was going to help people live better lives. It was going to change the world. And change was necessary with an aging population and consequential rising health care costs. Holmes raised capital from investors; first just a few million dollars, but soon after, hundreds of millions. By 2010 the company was valued at $1 billion. The company formed a partnership with Walgreens to distribute Edison through in-store Theranos Wellness Centers. They went public in 2014 and were valued at more than $9 billion, making its founder, Holmes, then only age 30, one of the youngest self-made multibillionaires in history. There was of course just one problem. The whole thing was a lie.

The technology behind the Edison box was a sham and so was the data used to support it. Edison did not work. It was later discovered that when prospective corporate partners visited Theranos' offices for a demonstration of the machine, the visitors were conveniently whisked away from the conference room between the time the blood was drawn and the results were shown. In the meantime, employees would remove the blood sample from the machine and take it down the hall to a real lab to perform the tests. Once they had the results, they would return to the conference room and program the Edison display to show the results. The data on the screen would show them the results of the tests, but, of course, it was the data equivalent of a sleight-of-hand magic trick. The magic was not the innovation of the technology, it was the deception using the data.

The story of Theranos is not the typical data duping we expect most people to encounter at work. But it is a powerful story about how deception by data can operate. Well-presented data provides credibility. Data can help support a decision or convince someone of another course of action. It can give the impression at times that things are "okay," leading its viewers to conclude no decisions or changes are required. Complacency by data, data that makes you feel okay with the current state of affairs, is a type of data duping. There are two aspects of the Theranos story that stand out. One is many of the people involved in the story wanted to believe the data were true. The other is they failed to ask the right questions to avoid being data duped. Perhaps they did not ask the questions because they wanted so much to believe Edison was working, or maybe it was because their unfamiliarity with data made them hesitant to ask a "numbers" question. Statistics may not be everyone's strength and it can be intimidating, especially in a room at Theranos where you are surrounded by scientists. Moreover, it may be difficult to know the right types of questions to ask. In this chapter, we promise not to intimidate or overwhelm you with statistics or mathematical theory. We will attempt to show you where data can show up in your business—any business—and ways in which you might commonly see it used in a deceptive manner. Also note, as we have written from this book's beginning, deceptive data are not always intended to be misleading. The creator of the data may believe they are showing the most accurate information, so this chapter is not about vilifying people, but more about making you aware and prepared. And being prepared means being able to ask appropriate questions about the data, foster discussion and evaluation, and improve upon how data are used in making the best decisions.

EVERYONE FACES DATA

You may not think of yourself as a data person. And unless you are in the analytics industry, you might not even dare refer to yourself as a data scientist. However, we are pretty certain everyone encounters data in their job. (If you have an unusual job where there is absolutely no data involved, please let us know.) As we have noted before, data and its application can be subtle, which gives us even more reason to become data-aware. Let's start with a few examples.

Building a road was once so simple. Someone would draw up some plans and surveyors would stake out the land, marking the boundaries and placement of the new road. Next large earth-moving equipment would follow those lines, removing earth or adding where necessary. They would drive along the markers, often strings tied to stakes, and make the straightest and finest road you could imagine. The data age has changed all that. Today most road-building equipment is connected to GPS and the internet. When a road builder needs to grade a road along a straight line to a specific elevation, it can be done with the utmost of precision.

The same goes for farmers. Tractors plough fields or harvest crops with the guidance of technology. Some experimental designs even turn over the task of driving to a fully automated tractor that can work the fields all day all by itself.[1] Other automated farm equipment can differentiate weeds from other plants and spray herbicides only in the places they are needed, saving 80 percent on costs. The stereotype of a farmer not knowing how to use data is quickly becoming a thing of the past. The change for the people operating this type of equipment is profound since running a million-dollar farming operation is now common and, just like running any other type of business, it involves data.

The coffee shop on the corner is no different. If you have visited one lately you might have paid for your latte using the popular Square point-of-sale system. Swiping your card on their stylish tablets or tapping your pay-and-go device not only has been a wonder for small businesses, it also provides them with a lot of data. Square recently gathered data from coffee shop sales and looked at a year's worth of information to help give some insights to store owners. Coffee shops can be a tough business with narrow margins and nearly 50 percent of them go out of business within five years.[2] They showed things like peak sales times (8:00 a.m., noon, and 2:00 p.m.), perhaps an obvious find, along with more subtle trends. They discovered nearly 1 in 20 customers returns the same day, and 1 in 10 returns on consecutive days. Insights like that can help shape loyalty programs as well as pricing.

We probably have all heard the reference that 99 percent of all businesses in the United States are small businesses with less than 500 employees. If you did, you were not duped. According to the Small Business Administration (SBA), you would also learn that many of those are much smaller. Businesses with fewer than 20 employees make up 89 percent of all businesses. Starting a small business as an individual can be a daunting task with a lot of uncertainties. This may be why nearly 50 percent of small businesses fail in their first five years. But many that survive might be able to hang their hat on good use of data. A study commissioned by Google and performed by Deloitte found small businesses that are more digitally focused had better results—two times more revenue per employee and six times more revenue growth than their less-data-focused peers.[3] The reason many small businesses do not apply more technology and data to their businesses is a bit surprising to us. The study indicated business owners were not limited by technology costs, skills, and tools—recall many of these are available for free or nearly free such as Google Analytics, Google Sheets, and a host of training on Coursera. The reason was the owners did not believe using digital tools—everything from online social media promotions and data tracking to forecasting sales in Excel—were relevant to their business and thus did not believe they would be effective. In our opinion and those at Deloitte, nothing could be further from the truth.

Admittedly we are data folks and we approach many tasks with a data orientation. As a small business there are a myriad of questions that can be answered with data, even before you open for business. Initial business plans, projecting sales, cost of goods, employee expenses, rent and other overhead, can be developed to help decide "what-if" the business is successful. Plans such as these can help set expectations for the often most financially difficult first year. It can even be helpful to peer into the future and see what is possible in two or three years and when your business will turn a profit.

Small businesses report 64 percent use social media to promote their business. Data from social media and a business's website also provide a lot of information about your small business. So how can all that data be used in a useful way, and not be deceptive on its own? First, you have to know the decisions you are making (the easy part) and the corresponding data that will let you know what to do (the harder part).

DECISIONS, DECISIONS

You use data to make decisions, right? Well, if you are running a small business or working for a large company, you likely find yourself as part of a decision eventually. You will see data presented by others or create some of your own. What is important is to know when data are manipulated toward a specific decision artificially, or by misuse of numbers, and when it can be trusted.

Sometimes we all have difficulty trying to describe something. When we are sampling a wonderful new dish with friends at a restaurant, we might say, "It has a hint of this or a dash of that." We are searching for the ingredients that are combined, creating this one flavor, and sometimes it is difficult to just say "it's coriander!" We just know it tastes great, but we do not know why. We do not know what went into it and how it all came together into this one great scrumptious thing. Data science is like that too. We have outcomes and have to work backwards towards the inputs, like ingredients, to figure out what causes something to be, or taste great. More often it is to find what is terrible and needs improving. Sagging sales, poor customer satisfaction surveys, declining quality in a factory's output. All of these can be viewed like our dish, and the ingredients are the data. But when you do not know all the data, you might just throw your arms up in the air and say, "just taste it," because we do not even know where to begin to name and estimate the ingredients. Solving business problems sometimes suffers from this same sort of problem, and a good place to begin is the scientific process.

Scientific Process

Somewhere around the time of middle school, you were likely introduced to the scientific process. Do not let the word "science" distract you. There is science in business. The scientific process is an elegant model for how to know when something is true and is applied to all sorts of science and business applications. It allows us to draw conclusions about an observation and its causes. The first step starts with an observation, such as a pot of boiling water and a problem statement— "what is causing the water to boil?" With a little research, we can form a hypothesis, a question we can test—"the water is boiling because of the heat under the pot." Next, we can perform tests on several pots of water, some with heat and some without. Finally, we will conclude the pots with heat (at least enough heat to allow the water to reach 212 degrees Fahrenheit) are the ones boiling. After a series of experiments and data collection, we will reach a definitive conclusion. If we do not, we repeat the process testing another hypothesis.

First the Essential Question

The scientific process sounds simple, right? It is simple as this modest example shows, but too often in more complex business situations, the concept of developing and testing an idea (hypothesis) of a root cause for a problem is overlooked. Especially when data are widely available, it is easy for data teams to want to quickly gather the numbers, substituting volume for understanding, and rush to present something. This can be terribly distracting. What if that something is not a cause of the problem, but more likely a symptom of it. There is a famous quote, often attributed to Albert Einstein, about the importance of understanding a problem *before* setting off to solve it.

> If I had only one hour to save the world, I would spend fifty-five minutes defining the problem, and only five minutes finding the solution.
> —Albert Einstein (maybe, but probably not)

It is a wonderful reminder no matter who wrote it, to first get an understanding of the problem you are solving and the factors that may be contributing to it. If you find yourself in a business situation and the presenters have not explained the connection between the data they are showing and the outcomes, then you may be vulnerable to a data dupe.

Metrics Dashboards

Business Metrics and Key Performance Indicator dashboards, the latter simply referred to as KPIs, may be the most suspicious examples of throwing numbers at a problem. They are the equivalent of flinging cooked spaghetti at the wall to see what sticks. Metrics *en masse* can be bewildering to managers trying to understand 1) what is the *real* problem and 2) what needs to be done about it and 3) how do all these numbers relate to the problem in the first place? Business metrics and KPIs should all come with a note indicating which ones are controllable and for the others, how much the outcomes would be impacted with one unit change up or down. Controllable might include the number of stores you own while interest rates are considered uncontrollable. The impacts would show the effect on sales with the increase or decrease in the number of stores or interest rates.

There are whole books written on business metrics and dashboards in particular. The intent of a dashboard is to present information effectively in a manner that is concise and relevant. Business metrics dashboards take their cues from real-life counterparts that you may see in your car or the cockpit of an airplane. Unlike a report, the dashboard presents information quickly, usually in a one-page or one-screen view. Like an airplane dashboard, the more important information is front and center and easy to interpret. Data visualizations are optimized to provide the best view of the information—uncluttered and meaningful to the decisions the user is tasked with making. In an airplane, these might be speed, heading, and altitude. In a business, it could be sales, revenue, and expenses.

Business metrics dashboards that are not well designed can lead you astray and dupe you simply by providing *too much* data. Too much data means data dumping and is a form of data duping because without a well-defined purpose for the dashboard, a plethora of different numbers forces the reader to do the equivalent of a detailed analysis in their head. That's an impossible task where the reader is attempting to draw connections between the data points. You might observe sales increase following a change in advertising, or perhaps it was related to a change in pricing strategy or maybe a broad bump from an upturn in the economy. There is no way to reasonably know simply by looking at it. After all, making numbers-based associations like a regression analysis is exactly what computers are good at doing rather than a manager staring at a metrics dashboard. Again, when multiple factors are presented in a dashboard, you should be asking "why?" There should be clear relevance and context. For example, presenting the percentage of sales compared to a goal is relevant to the sales manager, but not to the operations manager responsible for the factory. Context should indicate if numbers are good or bad and which way they are trending over time. Sales that are 10 percent higher than last year might sound great, but not so if they are below the goals for the current year. Just like designing the overall dashboard requires a specific purpose, customizing it for certain job roles, and this is the important part, *related to the decisions they make*, can yield a more effective dashboard. If all of that does not make much sense, then give the metrics dashboard you are looking at this quick test. Look at it for 10 seconds. Are things good or bad? Will things be good or bad in the future and why? Are there any problems that need to be addressed? What things can you do to change the outcomes and solve problems? If you cannot at least answer most of these questions, then you are being data duped and need to consider what metrics need to be removed and which need to be added.

But, Why?

How do you identify the important metrics and toss away the others? It gets back to asking more questions and creating testable ideas (hypothesis testing). If you are running a business and considering expanding with more locations, the number of stores may be relevant. Will sales increase with the number of stores and by how much? This should lead to other questions, such as how important is the location? What is the cost of expected rent? Will additional advertising be needed in the new market? And so on.

A good exercise for identifying root causes is called Five-Whys. It was born out of the manufacturing era by Toyota Motors and its founder Sakichi Toyoda in the 1930s and grew popular in the 1970s, spreading to other industries. Its premise was to encourage managers to ask more questions about the problems they encounter (how convenient in a book about questions!). The limit of five questions is not a rule but a guide as it may take more or fewer questions to get to the root cause of a problem. Toyoda believed until you are working on fixing the root causes then you are only treating the symptoms, those intermediary problems, without fixing the bigger issues. Think of the Five-Whys exercise as reminiscent of a curious child, with what may feel like a relentless series of why questions. Children and their curiosity are very good at getting to the root cause of things. Consider a business example that starts with a decrease in customer satisfaction:

Customer satisfaction is decreasing. **Why?**
Our product deliveries took longer than expected and were late. **Why?**
There was a shortage in manufacturing. **Why?**
Demand increased and all equipment was running at 100 percent capacity when a critical machine failed. **Why?**
The machine failed because it was operating beyond its regular maintenance date. **Why?**
We do not have surplus machines allowing for some to be offline to perform maintenance.
Root Cause: Machines cannot be maintained because there is no surplus manufacturing capacity.
Solution: Acquire additional machines to allowing the factory to properly maintain production machines. This will improve delivery times and customer satisfaction.

As Einstein may have pondered, these are exercises in patience and thoughtful design. They are aimed at understanding and defining the essential question. The essential question is that one thing that we really need to understand the problem. Without an essential question properly defined and answered, the recipient of our Five-Whys who simply stopped at the first one would be busy redesigning the routes of their delivery drivers, and our managers would still be staring at their metrics dashboards trying to do math in their head. Meanwhile, their competition, free of overwhelming numbers and avoiding data dupes, are winning the day.

THE SPREADSHEET—FRIEND OR FOE

If someone was going to be deceived by data, we think a spreadsheet might be a good place to start. Spreadsheets are easy to make and can create an overwhelming amount of data. Spreadsheets like Excel are also incredibly powerful tools. They can perform advanced calculations, manage large amounts of data, and even have a built-in programming language called VBA. Presenting a spreadsheet in a business meeting also provides an amount of credibility beyond what a typical PowerPoint slide can do. The challenging part is how to navigate the information you are shown and make sure you are making the right decisions based on the data.

The first item to be wary about are the calculations. Each non-empty cell will have either data or a formula, and some of those formulas can be confusing to follow, with many links to other cells being possible. Other calculations may just be wrong. If you have worked with a spreadsheet, you might know how easy it is to make a mistake, sum up the wrong column or omit part of a formula. In 2012 a simple input error caused the London Olympics to sell 10,000 seats they did not have.[4] Eastman Kodak had a similar problem in 2005. During a restructuring of the company an input error related to layoff costs resulted in a $9 million error. It was significant enough the company had to restate two quarters of earnings and resulted in an ambiguous and most memorable statement about the error. They wrote, "[there was] an internal control deficiency that constitutes a material weakness that impacted the accounting." In other words, someone accidentally input too many zeros into the spreadsheet,[5] but that was clear, right? Even the explanations of a spreadsheet error can be confusing (see figure 7.1).

Human errors in a spreadsheet are common, especially when data are not drawn from a database connection. This type of unintentional data duping is more about the complexity of the numbers than their

TIP

Look Beyond the "Blip"

When you are presented with a big grid of numbers in a spreadsheet, like monthly sales projections for all of your products, ask that it be visualized. Our human tendency is to scan for the biggest "blip" or outlier and then, perhaps unnecessarily begin focusing on just that. Maybe it's an error, maybe it was just a sales promotion. Either way it might just be a distraction to other things you should be looking at. Ask yourself what is the *next question* to answer if you had not seen that "blip"? Of course, get an answer to that "blip", just do not let be the only thing you do.

Figure 7.1. **Seeing those initial outliers in spreadsheets can cause you to overlook the real information.**

conclusions. If you are presented with a spreadsheet, keep this in mind. What you are looking at is a mosaic of both computational formulas and human input. Knowing how much of the latter is weighing in on your decisions will increase the uncertainty. There are a wide range of opinions about how many spreadsheets actually have errors with little insight as to whether those errors are cosmetic or fatal toward their purpose.[6] If the decisions you are making have high costs or impact on your business, you should consider the data defense strategies we mentioned earlier.

What is the source of the information and is there any possible bias from the people who created it—in other words, is there a motivation towards a particular decision? We are not being judgmental of people who create spreadsheets, but unconscious bias can be a real factor when the spreadsheet is created with a particular decision or end already in sight. The other data defense strategy is "too good to be true." The auditing tools within spreadsheets for finding errors have improved a lot since Excel was released nearly 35 years ago. Formulas can be inspected and traced to other dependent or precedent cells. The data connections feature in Excel can also show what other data sources might be used in the calculations. However, as the person reviewing the spreadsheet, you have to ask, "Is this possible?" A too-good-to-be-true spreadsheet might show an overly optimist sales growth, or a disproportionate ratio of expenses to revenues that are higher or lower than expected. Any of these should lead you to check the work and ask further questions.

Another point about spreadsheets is they only do exactly what they are told to do, for better or worse, and there are some limits. Here are a couple of examples we call "3-2-1" errors. Open up Excel or Google

Sheets and enter =(0.3-0.2-0.1). Logically it should result in zero. Using Google Sheets, you might have to expand the number of decimals displayed, and eventually, you will see it is not quite zero. In Excel, it will show -2.77556E-17, the E-17 indicating 17 decimals before 2.77, a rather small number, but still *not* zero.

Here is another. Enter =111111111*111111111, roughly 111 million squared. You should get a nice symmetrical-looking number that goes up and down from one to nine as 12345678987654321. But instead, you see 12345678987654300, which is incorrect. The "321" in this sequence at the end is simply trimmed to "300." Where did the "21" go?

There is an explanation for both errors, and it has to do with how computers store numbers. All computers use the binary number system because down at their roots in their microprocessors all those tiny transistors can be signaled only to be on or off, or rather in a state of a one or a zero. The number 2020 for example is stored at 11111100100, and the limitation according to Microsoft and the official standard-setting group, the Institute of Electrical and Electronics Engineers (IEEE), rule IEEE-754 says the maximum number allowed is 15 digits. Our number 12345678987654321 has 17 digits so the last two are dropped and replaced with zeros. There are workarounds for this including writing custom functions using the embedded VBA programming language, and unless you are astrophysics working with really large numbers this generally will not be a problem. The closest star to Earth, Alpha Centauri, is only 4.3 light-years away, or 25,454,000,000,000. This requires just 14 digits and is easily calculated in Excel. In fact, we could make calculations up to 100 light-years without a problem. What is a problem though is when this error occurs in a spreadsheet there is no notification, no warning that says "Hey, this is a very big number beyond my capability to store in memory, so I just trimmed off a few digits, okay?" That would be a really helpful message! Instead, Excel and Google Sheets just does as instructed and carries on, which can leave its users a little vulnerable to a data dupe and maybe heading towards the wrong star.

Math is great but presenting it in a fancy way like in a spreadsheet might lend it too much prestige. No one wants to question the math and look silly doing so. The math per se may not be wrong—two plus two always equals four—but how it is used could be inappropriate. Recall in an earlier chapter how convenient correlations can sometimes present as real-looking relations but are only a coincidence of numbers. And indeed, the math was correct but it was still wrong because of how it was applied. It is a case of the right math in the wrong place.

Often this type of misapplication shows up as inputs to spreadsheets, usually based on past trends. The creator of a spreadsheet looks back at recent history of some variable like sales and applies a growth rate of those past results and projects them into the future. It sounds reasonable, right? Except simply taking those numbers and extending them into the future does not tell us much about *why* our past sales were growing and what was the root cause for their fantastic growth. Small business owners may already be wary of this approach to forecasting future sales. Sales in a small business can start slow. As customers discover the product, sales can grow at exponential rates. Will these sales keep growing or is it just a result of the initial start-up of the business? Is it reasonable to expect this rate of growth to continue forever?

Not long ago, people had a similar concern about the growth of the human population. The world population was relatively small throughout most of history and did not grow much until the late 1700s. Population growth barely existed before 1700, nudging along at less than 1 percent each year. The world population finally reached 1 billion people around 1800 when the growth rate gradually began to increase. Beginning in the 1950s the rate of growth took another upward turn and the world population was growing faster than a rocket, passing 3 billion by 1960. Fueled by post–World War II developing economies, education, lower child mortality, and improving health care, the population crossed 4 billion in 1974 and 5 billion by 1987. It seemed there was no end in sight as the population grew and grew. Many noted the growth was occurring more quickly in economically poor countries, with their population projected to double every 24 years. In 1992 researcher Göran Ohlin, attempting to bring a more rational view to the data, wrote about the concerns of world population growth:

> the argumentation about population growth has been unduly dominated by mere numbers and by scary extrapolations based on compound interest, which easily baffles unaided common sense. In addition, some very primitive economic reasoning tends to stick to population alarmism. If India is so poor today, the reasoning might go, how is it going to feed and support a population that will be twice as large in the next generation and four times as large in the generation after that? There are also many

variations on the theme that "we" will be swamped by "them." "They" .
. . threaten to displace the peoples of the rich and industrialized countries
from their already shrinking role in the world.[7]

Clearly, those who fell into this belief about how population growth
was an endless upward straight line based on the (recent) trend were
data duped. Their fears and concerns as citizens and governments of
slower-growing and more economically rich nations were reinforced
by data. How could that be wrong? After all, they had seen how much
the population was growing both in terms of total numbers and growth
rate. They had seen an increase of more than 1 billion people in just
the past 10 years.[8] Ohlin goes on to berate these protagonists for this
insensitive and inhuman view of struggling nations and the inherent
responsibility of others to provide them aid, even if these projected
numbers were true. However, as Ohlin notes, they had been "unduly
dominated by *mere numbers* and scary extrapolations." The numbers
made them do it! Otherwise, rational people would comprehend these
numbers rationally. But they didn't. There are two important lessons in
this example that you will also likely encounter in the business world.
First, when we are presented with data that reinforces our preexist-
ing beliefs, we are more likely to simply accept it as true. That bears
repeating. When you see numbers that you *think* are true you stop
thinking.

People who believed the world population was growing continu-
ously did not question numbers that also supported their same belief.
In a way, their skepticism of numbers was momentarily suspended.
Their brains were turned off to questioning the rationale of using those
numbers as inputs to project a future state of the world. They simply
accepted it and moved on to what they believed was the real problem,
that is, how to support all these billions of people that soon would overwhelm the planet. How often does this occur in a business setting? How
many times does the incorrect conclusion based on a spreadsheet shift
work and focus on a problem that is not really a problem at all, only to
later discover time and resources have been wasted?

The second lesson is the flaw of using a trend as an input to project
the future without understanding thoroughly the root cause and drivers
of the input. To take the recent population growth rates and project end-
lessly into the future was unrealistic. It does not matter if this was done
in a formal spreadsheet or the minds of the citizenry who reacted to this

belief. The input of a trend requires rational purpose and explanation. Just like the swooning first-year sales of a new start-up business are unlikely to continue forever, so are many other trends. Digging deeper into the recent trends of population growth would facilitate discovery and understanding. The drivers such as economic growth and a stark decrease in child mortality were among those factors leading to increasing populations. However, increasing survival rates does not mean, for example, that parents will continue to have more and more children. The trend is not endless, and, in fact, the average number of children per family begins to decline. After understanding these numbers and others that contribute to world population growth, the population is expected to plateau in the mid-2050s at around 10 billion people. For all of those governments and concerned people that Ohlin wrote about in 1992, those who feared the world was on the edge of a population crisis, they may find they wasted a lot of time and energy because they were data duped.

SCENARIOS AND SIMULATIONS

In 1946, Stan Ulam and John von Neumann were working at the Los Alamos National Laboratory on the highly secretive Manhattan Project that would eventually lead to the world's first nuclear weapon. Although they did not have the benefit of a modern spreadsheet, they had a similar need to estimate a problem. They were trying to understand how a particular situation of a neutron would work and, of course, doing a real live experiment would be very dangerous. They had a deterministic understanding of how the physics worked but needed a method to estimate uncertainties.

While playing a card game of solitaire, Ulam considered how the randomness of the cards might help him solve the problem. What they eventually developed was a framework to insert randomness into a simulation where there were some known boundaries. For example, if you drop a ball on the floor from a certain height you can predict how it will bounce and how long before it comes to rest. Drop 20 balls all at the same time and the interaction of each ball hitting another can make predicting when one in particular comes to rest more difficult. There is a random chance it will be hit by another ball and interfere with where it falls next, either speeding up or slowing down how long it takes to come to a stop. If you knew the probability of one ball hitting another, perhaps from observation in another experiment, then you could assign this probability to this event in a simulation. You could repeat this process for several variables and assign each a probability. However, without a simulation with all these variables and their respective probabilities, it is difficult to know the most likely end result of all those things put together. In addition to probabilities, they also needed a reliable method to generate random numbers. Just like in a card game there is a 1 in 52 chance of flipping over a specific card, or a 1 in 13 chance if you do not care about the suit. When running a simulation, you would need to pick random numbers from 1 to 13 to simulate a particular card value. The task of generating a truly random number is quite difficult for a computer and still remains a challenge for data scientists. However, in 1946 being able to generate a *nearly random number* was good enough. They devised a method to come up with a set of random numbers and began inserting them into repeated simulations.[9] By observing the number of successful outcomes, they could then derive a reasonable statistical estimate of how the neutrons would behave. Since their work was secretive, they created a code name for their method called Monte Carlo, based on the famed casino in Monaco we mentioned earlier, where random chance made winners or losers in games like blackjack and roulette.

Ulam and von Neumann's work on simulation using random inputs provided a methodology that has been applied to many other areas such as weather predictions and financial modeling. It is also very useful in spreadsheets where there are future projections and there is some amount of quantifiable uncertainty.

A quantifiable uncertainty in a spreadsheet might be the future projected sales of your product. Perhaps you *know* from past years that your sales ranged from $50,000 to $70,000 per month. Some months are better than others and sometimes that is just the way it goes in business. Now, suppose you are planning the future of your business, what do you input for projected sales? A common approach is to just take an average, $60,000 in this case. And this would be a fair approach, but what if you did this for all your projections—the cost of goods, rent, employee expenses, advertising, shipping, equipment maintained, and on and on? Certainly, some of these inputs are likely fixed, rent for example does not vary randomly, but others do not. Equipment for example can break down and require replacement or repair. And within a range, sales can be variable too. Maybe you sell ice cream and hot sunny days are a boost. Or you are a roofing installer and rainy days mean lost revenue. To account for randomness in your spreadsheets, you can introduce a Monte Carlo simulation.

A Monte Carlo simulation is like being able to run your business repeatedly under different scenarios to see what happens to your financial results. Spreadsheets are an excellent tool for this task. Basically, you create the same spreadsheet as before, but instead of using average sales, you input a range of sales perhaps $50,000 to $70,000 and then compute a series of simulations, maybe a thousand times. After the simulation, you observe, simply by counting, how often you had a good outcome (made a profit) or a bad outcome (financial loss). Rather than just relying on an average you can see how often a range of sales can either be good or bad for your business and in effect the probability of your success, something a typical spreadsheet does not show. Creating a Monte Carlo simulation in a spreadsheet is not difficult to get started. There are some built-in features in Excel to ease the task. No matter if you are creating the spreadsheet yourself or being presented with one showing future projections, it may be well worth asking about a Monte Carlo simulation of the uncertain inputs. The simulation can lead to a discussion of scenarios and uncertainty regarding the outputs from the model you care about.

When you run a simulation, you get a better understanding of how the uncertainty of estimates used as inputs in your spreadsheet can impact the outputs from your model. Ideally, you will have some information from the past pointing you in the right direction. Perhaps you know cloudy days affect ice cream sales by 20 percent or small changes in interest rates affecting your cost to borrow can make or break your bottom line. With this type of information, you can create different scenarios and look at the resulting outcomes. Rather than using averages for input variables to create one spreadsheet, consider three. These three can be the best case/worst case/typical case scenarios. For example, if you are thinking about starting a small business and you have a spreadsheet where you are trying to look forward at how it will do, look at the very worst scenario. Assume you have dreadful sales and unexpected expenses. Where everything that could go wrong does. Your suppliers charge more than expected. Your bank approved the loan, but at a higher interest rate. Your equipment breaks down and needs repair in the first six months. If you are still profitable when many things go wrong, well, you have a really great business idea. Then look at the other side of things. What if . . . everything is great? Sales are more than expected. You win a surprise contract with a large company, ensuring steady sales each month. People love your product so much you do not need to spend much on advertising. How does your business look then? Probably pretty good. Do the same for a thousand different scenarios, where some things go well and others do not. After going through each of these scenarios you will get a better understanding of what might make or break your business plan, and maybe even some insight into what items require more of your attention as your business grows. In doing this exercise you have reduced your chances of being data duped by a rosy set of numbers that may be based on past averages. You can apply these same concepts in a corporate setting when you are viewing a presentation. Ask about how certain outcomes like sales, revenue, and variable expenses fluctuate to changes in inputs—how sensitive are they when things change? Ask for a "scenario analysis" to show the best- and worst-case situations. We believe when multiple scenarios are presented and they still consistently point to a decision, well then, you are not being data duped. Failing to ask for this type of workup leaves you vulnerable to an optimistic, and unrealistic, view of the future. Let's look at a real-world example.

By the time John Weikle first founded Skybus Airlines in July 2003, passenger air travel was rebounding from the aftereffects of the 9/11 2001 hijacking and attacks. While air travel was up, the event and its aftershocks on the airline industry demonstrated there was a need for a more nimble airline business model that could encourage travel with lower-cost fares and decrease overall expenses. Skybus attracted notable investors such as Fidelity Investments and Morgan Stanley. Their business model was edgy. You were more likely to see flight attendants wearing T-shirts than a pantsuit and they were selling merchandise during the flights. In fact, the on-board gifts, snacks, and perfume sales were of a "hip" brand atmosphere and an important means to supplement ticket revenue. They marketed variable fares where the first few tickets on each flight were only $10. They were also among the first airlines to charge fees for checked baggage, a practice that today is commonplace. They even generated revenue by selling advertising on the inside and outside of their planes. They were innovative, adapting the practices of other European discount airlines and adding their own improvements. The airline industry was changing along with the economy that drove it, and they were creating an airline for the future. One thing they could not anticipate was the future costs of airline fuel. Or could they?

Passenger travel was indeed gradually improving, signaling customer demand for airline travel. But like many businesses, it takes time to get things going. When their business was first incorporated in 2003, fuel prices were $1.51 per gallon. Between the time their Department of Transportation (DOT) application was filed in early 2005 and approved in 2006, fuel prices increased from $1.83 to $2.74. This was important because an airline's most uncertain expense is fuel. Fuel accounts for 15 to 20 percent of total expenses. Unfortunately for Skybus, they were heading towards a worst-case scenario.[10] Skybus, which was striving to have the lowest costs per passenger mile in the industry, was suddenly looking at fuel costs that were more than 50 percent higher than when the business started. A year later they would begin their first flights while fuel costs were still climbing. By the time they collapsed in April 2008, having operated for barely 12 months, total fuel costs had increased by a staggering 228 percent! This is the type of situation that can benefit from Monte Carlo simulation and scenario analysis.

Fidelity Investments, Morgan Stanley, and other Skybus investors should have been wary of a business model that relied on expected costs aiming to be the lowest in the industry. Maybe they did, but the evidence suggests they did not. There was continued optimism as they approached their first day of flight operations. By October 2006, when fuel prices had already increased by 48 percent, they announced a long-term deal to purchase 62 new Airbus A319 aircraft. By April 2007, when fuel costs had more than doubled, they raised an additional $70 million in their second round of equity financing, bringing their total investments up to $160 million.[11] While fuel costs, their single most expensive item, were rising, they continued to move forward. Critics will say that no one can see the future and that the fuel costs at these prices were not sustainable. They would argue fuel costs would soon come back to normal. And they eventually did . . . in 2009. The point is, using techniques like scenario analysis and simulation might have given a better sense of Skybus's vulnerability by demonstrating the impact of growing costs and providing an estimation of the probability of the airline's success (good scenario with lower costs) or failure. Upon reporting their bankruptcy filing, an industry expert noted of Skybus, "Their fares and cost structure were out of line . . . Fuel prices rising is the type of thing you have to factor in when you do your risk analysis. If you didn't do that, then your management was very weak from the start." Maybe instead of $10 fares, they should have been $14 or $20 or somewhere in between? "What-if?" is a question all should have been asking to see in those spreadsheets.

LIONS, ELEPHANTS, MICE (OH MY?)

Should we be concerned about mice, at least figuratively?

This is a story about life in the middle. A story about mean versus median. If you know a little about statistics, you know the word mean is used interchangeably with average. Everyone wants to know about the average although no one aspires *for* the average. It goes like this. A client requests an analytics department to measure the "average" customer. Maybe it is based on the number of products the customer buys or demographics like the customer's income, home value, or age. Whatever it is, the analysis is often very focused on the "average" customer. Next, with the "average" customer in mind, the client starts to shape business plans around this mythical average customer. And often mythical is the right description, because the average customer rarely exists.

This parable may help illustrate the deception of averages. Members of a research zoo have traveled to the African safari to collect animals to study. They previously set up cages to capture the animals alive and are now returning to collect the animals and transport them back to their facility. The researchers are being assisted by a local expert who is very eager to see the success of the researchers he is hosting. Upon their return, he can see that their cargo trailer is heavily loaded as it weighs down on its supports. The expert smiles at their success. He says to the researchers that they must have been very successful as he looks at the weigh scale connected to the trailer and it shows a very larger number for the total weight of the cargo. It is an indication that many animals have been loaded onto the trailer. He queries the researchers about how many they have captured. He does some quick math in his head knowing very well the typical size of all the animals in his area of this wilderness and he concludes there must be many lions in the trailer. The researchers look sheepishly at each other. They have no lions. They were not successful in gathering many larger animals at all, only several hundred mice. And one elephant that needed a ride. The problem, of course, with the conclusion was the *average* weight.

The misuse or perhaps the misunderstanding of averages occurs all the time in the business world, especially when there are outliers or when the distributions of numbers are skewed one way or the other. Skewed distributions commonly occur in items we mentioned earlier such as savings and retirement accounts, home prices, and income. How might it also show up in a business case? Consider an example where the product you are making depends on the number of children in a family. The engineers want to know the requirements so turning to a survey average might be the first place to look. Recent surveys show the *average* number of children is 2.4 per household. And since the product is for people, it would be rounded simply to 2. But how many families have exactly two children? The answer is 53 percent. As a product designer, the engineers are building something that targets only about half of the potential market. In fact, the remaining group, those with three or more children, make up 43 percent of households with children. Given this insight, it may be time to ask the engineers to make a product that is more flexible to accommodate two or more children. Simply designing a product aimed at the (rounded down) average number would miss a lot of potential customers.

Sometimes designing for the average can have more serious consequences. In Australia, floods are the most dangerous and damaging natural disaster. Each year floods cost the country $377 million and sometimes have exceeded $5 billion, such as the flooding disaster in 2011. Engineers and scientists' approach to minimizing destruction from floods relies on complex weather forecasting that begins with a review of historical flood heights. Now as you can probably foresee, there are grave consequences to simply using the *average* flood levels and building barriers to that height. Even if flood heights were statistically normally distributed, it would mean that 50 percent of the time the flood level would be greater than the average. However, historical flood levels are not typical and not normally distributed. Most floods are small in any given year, but sometimes there are extremely large floods. Again, it is tempting to take an average level. Even when the distribution is skewed like this, the probability is much less than 50 percent that the next flood will exceed this height. There are a couple of challenges for the engineers that are equally useful for any business situation. In places like Australia and the United States, the historical records for floods may go back only 100 to 200 years. That may sound like a lot of time but making a statistical forecast on so few data points would be like guessing tomorrow's stock market based on the last 100 days' performance. It is a relatively small number of data points to make good estimates. If business decisions need to rely on historical data, having more is often better. The other challenge is how to estimate the skewed distribution's likelihood for rare circumstances to arrive at a reasonable measure for a 100-year event. The 100-year event is the common standard for decision-making entailed in making requirements for flood walls and building construction.[12] The solution is to apply statistics projecting the flood heights and providing ranges of likely outcomes. These give boundaries for the estimate and can aid in making a better decision. Likewise, in business if someone has provided a future projection, say sales, a good follow-up question would be to ask about the "90 percent confidence limits" of the projection. For example, sales are estimated to be $400,000 with a 90 percent confidence limit providing expected sales between $300,000 and $500,000. Knowing the boundaries of an estimate help with the decision. Would it be reasonable to go ahead with the decision knowing sales could be as low as $300,000? After all, we are 90 percent confident the sales will be between $300,000 and $500,000. Avoiding the temptation to rely solely on an average can make for more informed decisions. Just like the flood engineers, they will not design a barrier wall that is average. They will look at the range and the distribution of flood levels and maybe build the wall at a height such that there is only a 10 percent chance a flood will be higher than the wall.

Designing for the averages or making decisions based on averages seems reasonable. It is easy to calculate an average and generally easy to explain and communicate your decisions. This may explain why it is so widely used in business decisions. However, as we have demonstrated, averages alone can be deceptive, especially when presented without context, without ranges, or without an indication of their skewness towards extreme values. Designing for an average can result in a "one-size-fits-all" decision where you may later discover that one size fits none.

GET BACK TO WORK

As one of our former managers would say after we created what we thought was some brilliant piece of work, "that's good, now get back to work!" Her point was just because we could do some good analytics the task was not done. Knowing the numbers, the perils of misinterpretation were all fine and good, but it only really mattered when you could take that knowledge and do something constructive with it. Otherwise, it was (ahem) just academic. Applying analytic insight to influence a decision makes a product faster and cheaper or simply serves a better cup of coffee. Taking those insights into the real world is what truly matters.

KEY POINTS

- Every business is a data business. No matter if you are running a small business or working at a large company, data are inevitable. The more you can use it effectively, without being data duped, the better.
- Avoid putting the data before the question. Data are about making decisions. Before it is gathered make sure the right question is being asked. Use techniques such as the Five-Whys to find the essential question first.
- Spreadsheets are excellent tools to help just about any business problem, but they are also easily used as a tool of deception, sometimes even unknowingly by their creators. Spreadsheet models are always vulnerable to average inputs and unrealistic projections.
- Consider scenario analysis and Monte Carlo simulation to be better informed about the probability of a decision's success or failure.

All the Patterns We Shall See

How Our Beliefs and Bias Can Cloud Our Decisions

> Data and its exploration can be endless. We cannot know everything but sometimes everything we know is enough.
>
> —Anonymous

INTRODUCTION

In chapter 1 of this book, we wrote about the different types of data dupers—the unwitting, the devious, and the fallible. In this chapter, we will revisit these archetypes and explore more deeply into the fallible data duper and specifically the psychology of data deception and decision making. Why bother writing about the psychology and the mechanics of decision-making? The reason is quite simple. The more you know about how people make decisions and our innate human vulnerabilities, the better you are at identifying these types of "data missteps." We need to be aware of how things work, just like we showed in the chapter on data defense, so we can navigate data deception by recognizing our human nature and decision-making liabilities in others and, perhaps most importantly, in ourselves.

WHY WE ARRANGE THINGS

It is natural as humans to want to bring order to disorder. We all have a basic need to sort things—the dangerous from the safe, the known from the unknown—and react and make decisions accordingly. As a result, we tend to make the puzzle pieces of life fit even when they don't because the lack of explanation is unsettling to us.

What would happen if you randomly walked through a crowd and dropped a stack of papers? Would people just walk on by and leave the disheveled papers on the ground? Perhaps due to the kindness of strangers, some would stop and help you collect and rearrange your papers. Perhaps, however, the disarray in the first place would be the motivation to put things back in order. Bringing order and understanding in contrast to chaos and unanswered questions helps us cope with unknowns in our world. Whether this is explaining the *real reason* your favorite sports team lost the big game last weekend or the (very factual) timing of the next occurrence of Halley's Comet, there is solace in knowing information and the data that drives it. However, if you knew all the data that caused your sports team to be successful, from the health of the players, the speed of the ball, and the temperature of the day, would you feel much angst when they lost a particular game? Probably not if the outcome was truly formulaic. Also, it would not be much fun to watch. Perhaps surprisingly to data folks, there is joy and anticipation from the random events that sometimes occur in sports: a tipped football pass intended for another; the golf ball that barely rolls into the cup thanks to a slight breeze; or that incredible half-court basketball shot at the buzzer. These extraordinary events are beyond the typical data for sports, but their possibility does make it more enjoyable to watch. Contrast the otherwise predictability of sports with predicting weather and you might get a different reaction. No one says "Hey, this is great. Look at all the unforecasted rain we are getting!" unless, of course, you are a farmer.

Ancient people around the world had various explanations for natural events. The Greeks explained the rising and setting of the sun as a result of Helios, who would pull the sun across the sky in his chariot starting in the morning and ending in the evening. The Sami tribes indigenous to Finland and northern Russia described the Northern Lights as the souls of their dead. And the ancient people of Hawaii described active volcanos that burned forests and changed landscapes with lava flows as a result of a love triangle involving Pele, their goddess of Kilauea. In modern times, we have acquired enough data and science to more adequately explain these occurrences, but imagine for a moment the fear and uncertainty people must have felt without knowing the causes of these events. How reassuring it must have been to know some of these dreadful things could be predictable or perhaps controlled by simply respecting the dead or being faithful to your lover. We seek certainty from uncertainty and resolution from chaos. Bringing a sense of understanding to something we do not know how to interpret is an age-old device. In a way, we require some equilibrium when data and events are out of balance with our understanding. To reconcile these situations, often the data are interpreted. A thread of explanation is woven to match the known data points and a conclusion is drawn. In ancient times, the data of the changing position of the sun in the sky is paired with a known means of moving something with a chariot. The story fits. The problem of course is even with data, the observable motion of the sun, the story can still be wrong. And stories that misuse data are the plague of data duping.

We have a strong inclination to see patterns and connections and the reason for that is we humans find comfort in knowing the predictability of things and conversely feel unease when we cannot. A flash of lightning during a storm that is followed by the sound of thunder shaking our homes is predictable. Replace that same experience absent the storm and lightning and it may be cause for panic. Science of course is a great method for trying to understand the connection between events, that is, the patterns and their outcomes. Due to science, we know that lightning is the electrostatic discharge of energy and it results in the thundering sound as it travels through the air. But before science and the ability to collect data, people still had an explanation by piecing together things they knew and observed.[1] And we still do this today. If you have ever gazed at clouds and saw whimsical characters in the sky or stared at a Rorschach Ink Blot and said "I think I see a pattern, perhaps two people having a conversation," then you know what this is like. We arrange objects and ideas alike. Arranging ideas is how we formulate theories and answer questions about the unknown.

A MIND'S EYE

When we gather information, no matter if it is spoken or data on a page, our mind tends to immediately organize it in a way for us to better understand. One way our mind arranges data is by conceptualizing it as an image. If the old saying that a picture is worth a thousand words is true, then maybe the opposite is true too—a concise way to store a thousand words is with a picture in our mind.

When we describe a sunset near a tropical island, with the sun slowing sinking into the horizon, simmering its reflection off the ocean while creating a radiating amber glow across the sky, what do you think of? Are you processing data points or do you already have an image in your mind? Where a computer may assimilate the description of the sunset into pixels and assign numbers to each, you are more likely to create a visual. It is simply easier for us to comprehend this way. It is not a trick of the mind but more of a tool. Learning studies have shown that people learn better when shown information in multiple formats, written, spoken, and visually (demonstrated or through imagery).[2] It is an example of the efficiency of how our brain works and at times a liability. What is also interesting is our mind's ability to form these images even when we have not seen the actual object described.

Consider this example of a graph you have never seen before. The graph represents the difference in annual income based on education level. Moving from left to right the height of each bar represents the amount of income. As education increases towards the right, the bars are larger. Each bar is taller because as education increases so does income. At the bottom of each bar is a label for the education level. There is one exception for the last bar for a doctorate degree, which is slightly lower than the previous bar for professional degree (medical doctor, lawyer). The bars are blue with numbers at the top each. The first bar is $27,040 and the last one is $90,636. As we were going through each step describing the graph, the image has probably been building in your mind piece by piece. In fact, if we asked you to sketch the chart you likely would create something pretty close to the actual graph. And that is because your mind has been processing the data visually. Some of you may have remembered the numbers, but all of you would likely better remember the pattern, the general outline of what it looked like. How did you do?

Your mind works in this example a bit like the video game Tetris when it is fitting the "blocks" into place to complete the puzzle, thereby making it less of a puzzle and more of a coherent item. As we described the graph, your mind was busy visualizing. The mind is good at filling in missing pieces or at least anticipating what fits. This works with words too. A few years ago, there was a meme circulating that helps demonstrate this effect. Try reading the paragraph below as fast as you can.

Unisg the icndeblire pweor of the hmuan mnid, aocdcrnig to rseecrah at Cmabrigde Uinervtisy, it dseno't mttaer in waht oderr the lterets in a wrod are, the olny irpoamtnt tihng is taht the frsit and lsat ltteer be in the rhgit pclae. The rset can be a taotl mses and you can sitll raed it whoutit a pboerlm. Tihs is bucseae the huamn mnid deos not raed ervey ltteer by istlef, but the wrod as a wlohe. Aaznmig![3]

How did you do? Even though more than half of the words were misspelled, you probably were able to read it without too much trouble. It demonstrates that when we *want* to understand something, we can easily just fill in the gaps or missing information. Would we do the same with data, finding numbers to support what we expect to see? It is quite likely we do.

Another example of how our mind works to fill in missing information is demonstrated with 3D stereograms. A stereogram, sometimes called a magic eye picture, is a two-dimensional image that interweaves another image, thereby making it appear as a three-dimensional object. Relaxing your focus on the picture allows your brain to make connections within the two overlayed images. Of course, the picture is still two-dimensional, but once your mind's eye fills in the missing information it can see something that was not there before.

The point of all of this is when you have previous reference points, like the correct spelling of a word, you also have an expectation of what something means or looks like *before* you fully see it. Your mind uses shortcuts and jumps forward to a conclusion. This is exactly what can occur with data deception. Our mind will formulate an idea before the data are complete and jump forward to that "a-ha" moment when you (falsely) believe you know everything the data were meant to tell you. No need to look any further.

THE POWER OF STORYTELLING

Author and University of Jerusalem professor Yuval Noah Harari wrote in his book *Sapiens: A Brief History of Humankind* that humans are great storytellers. Humans are highly social beings. The social bonds that often keep our societies together are done so through *fictional* stories. Some of these are so magical, so inspiring they can call men to arms to defend against their neighboring village, collaborate to raise a barn or form a company, and even convince others to pause and set their work aside and show up at a place of worship. Harari adds that unlike any other animals on Earth, humans are unique in their ability to create and socialize these stories and produce references to things that do not exist such as a mythical entity (a business or a god) or a future imaged state (think the Garden of Eden or Shangri-la[4]). Animals, in contrast, rely on the facts in the moment, while humans can have their decisions swayed by other imagined outcomes. After all, *Data Duped* is a book about decision-making and the sometimes-bewildering task of processing information. A story in the form of an eloquent speech *can* make that task easier. Imagine this example of the famed battlefield speech of Scotsman William Wallace at the battle of Stirling Bridge in 1297, as portrayed by Mel Gibson in the 1995 film *Braveheart*.[5] In response to the unmotivated army, which shouts "[we are going] home, the English are too many," Wallace says:

> I am William Wallace, and I see a whole army of my countrymen, here, in defiance of Tyranny. You've come to fight as free men, and free men you are. What will you do without freedom? Run, and you'll live . . . at least a while. And dying in your beds, many years from now, would you be willing to trade all the days, from this day to that, for one chance—just one chance—to come back here and tell our enemies that they may take our lives but they'll never take our freedom!

It is a good speech. It may not be historically verbatim, but it is representative of how stories can paint a picture of the future and the need to make an immediate decision in the present. But notice there are no data, no numbers or other information about the situation. This is almost the complete opposite of being data duped, where the urging of the decision is done *without* numbers. There is an emotional appeal and an acceptance of the future outcome (none question their certain freedom in the future). If you were a data-minded soldier in the crowd, you might have raised your hand and asked some follow-up questions, like how many soldiers do we have and how many are on the English side? (The Scots were vastly outnumbered with reports of their 5,000 versus 10,000 English soldiers including 2,000 calvary.) As Harari asserts, stories are the fabric that hold together our societal groups and we have an evolutionary predisposition to creating and following them and to that extent being influenced by them. Although this example is a dramatic scene crafted in Hollywood, it may not be too far off from situations you may experience. Sprinkle in a few data points for context and credibility and you may see these embellished data stories in advertising, in the news, or at work.

Harari's work and others make a case that we have a predisposition for fictionalizing information baked into our DNA. What he describes as making rumors and spreading gossip is part of a type of survival mechanism. And as Darwin has taught us, survival instincts get reinforced through the generations. We are storytellers and we are getting better at it. Does that mean we have a certain positive response center in our brains to fictional stories, rumors, and gossip and a disdain for data and facts? Maybe if it means we can survive relationships and dodge those who conspire against us. Does it mean that we are susceptible to misinformation and that being data duped is inevitable? Probably not. Just like other forms of enlightenment, we need to continue to strive to overcome the quirks of our evolutionary past, perhaps overcome our internal beliefs in the face of new information, and learn to check our bias at the door. But first, let's take a look at some of those common biases.

ME, BIASED? THE UNKNOWN BIAS PROBLEM

Making decisions can be a complex task at least for things beyond our basic needs like knowing when to breathe. And how people make decisions will inevitably vary from person to person. Some, when faced with a big decision, may create a pros and cons list. Others may approach the task by creating scenarios of multiple possible outcomes and then choose the best and most likely result. These methods are, well, very *methodical*, but not all decision-making follows logical paths or decision trees that thrive on data and information. Sometimes we just make an in-the-moment decision. When it is a type of decision that we have made in the past repeatedly, then it becomes easier. We no longer need to weigh the pros and cons, the inputs and outputs, separating the data from the supposition, since our experience has already educated us on what to expect and what to do. Some might call this a so-called "knee-jerk" reflex decision or a "gut-decision," but in reality, sometimes those are just as informed as some other decision that is more thought out. Other times, however, relying on our preconceived beliefs subjects us to our own bias or the bias of others, which can negatively impact our decision-making.

A cognitive bias is a mechanism, a shortcut for our brain that we sometimes use when making decisions. It can be a quick judgment reinforced by our prior experiences and is born out of our need for mind efficiency. Our brain is like a complex multitasking computer and it has evolved to be efficient in learning. When it has learned something in the past it is quick to replace the thinking process with a shortcut and arrive at an answer, sometimes without giving it much thought. For those of you who are data scientists it is the computer equivalent of using a subroutine or a function from a code library. Rather than working through all that code to answer a question, the function conveniently and quickly provides the answer. If you learned your multiplication tables in grade school, you might already experience this when someone says "what is five times eight?" In your mind, you associate the phrase immediately with 40 instead of working out the math of adding five together eight times. The same is true for words when you are reading this sentence. You do not take the time to sound out each word letter by letter since you know from recognition the meaning of each and can quickly understand the meaning not only of the words but the whole sentence. These cognitive shortcuts are learned from our experiences and can impact our decision-making in the face of data. Our ability to understand and recognize biases in ourselves and others will not only give us insight into how decisions are made but also help us avoid being data duped.

There are many cognitive biases and we will not attempt to cover all of them here. Rather we will focus on a few that are common to data deception.

Confirmation Bias

"Find something that proves this thing I already know!"

This may be the most common bias encountered in the world of data duping. A good data-driven decision should evaluate all the best information and leave confirmation bias at the door. As we referenced before, following the scientific method and treating data properly leads to the best results. When the process is disrupted, perhaps when data are filtered toward a preconceived idea or there is a lack of experimental design, then the risk of data deception increases. In individuals, it is the tendency to favor information that already aligns with their current beliefs. It is often the underlying mechanism of myths and legends. Find a four-leaf clover and then have something lucky happen? How many times would this have to happen to you or you heard it happened to others before you'd believe four-leaf-clovers are a sign of good luck? Perhaps this is an easy one to dismiss since we know four-leaf clovers are unrelated to random luck . . . right?[6] But it is exactly this type of narrow association that leads to the false beliefs such as the incorrect association of autism and vaccines or that colder weather causes the seasonal flu. Both are incorrect, but we naturally tend to associate unrelated data points to make conclusions. Again, we see this need to bring understanding to misunderstanding. Many of our false beliefs are a result of confirmation bias as we attempt to confirm one thing, we associate with another, while often ignoring contradictory information.

In 1960, English psychologist Peter Wason set out to explore and disprove earlier ideas that humans reasoned by logical analysis.[7] How could someone reason in an illogical and irrational way? To answer this, Wason created the "2-4-6 Problem" as an experiment to see how participants would perform rationally or irrationally when solving it. Participants were given the sequence of numbers and told there was a rule in play that determined the numbers. They were also given multiple opportunities to test the rule with the person administrating the experiment. What Wason learned was most participants latched on to an initial and incorrect belief of the rule—the numbers were two more in value than the previous. More importantly, they used their attempts to solve the rule with the administrator to further confirm their initial belief, rather than test an alternate belief. Their bias for falsely believing they knew the correct answer had already kicked in. For example, they would submit 8-10-12, to which they were told it followed the same rule but was not the rule in play. The rule in play was *any* three numbers in ascending order. So, for example, 10, 23, and 109 satisfies the rule just as well as the example given. Nearly a third of his participants seemed

Chapter 8

to be trying to confirm their initial conclusions rather than rationally seeking the correct understanding of the rule. From this experiment and others that he later developed, he coined the phrase "confirmation bias." By the way, if we had had many more examples, the rule inlay would have become more obvious.

Confirmation bias is something you have likely seen in the news but maybe did not recognize. It is a process in motion behind many poor business decisions. Consider this scenario. A manager for a new product is eager to introduce it to the marketplace but lacks information about how it should be positioned in the advertising campaign, let alone how much the price should be that would attract the most customers. The manager doesn't just lack perfect information, but like many times in business situations with something new, they lack information period. There is pressure from management to move quickly with the product launch and start to make money to recover the cost of development. The timing is critical and cannot be soon enough. (Realistically when we speak with business leaders, this is more often the rule than the exception—most projects are behind schedule and over budget. This in itself is a whole other decision-making conundrum.)

What does the manager do? The manager has likely already formulated a few ideas about the pricing strategy and the advertising "pitch" to customers. They just need a little bit of data . . . to confirm what they already know. In a rush to the finish line of the product launch, they limit the analysis of the "ideal" customer segments to just a few. The price points are largely predetermined because there are few opportunities to do in-market "test-and-learn" events. They make their decisions, create the advertising campaigns, launch the product, and with all the buildup, the anticipation, the wonderful expected return on investment turns into a failure. It will likely take a few years of data analysis to unwind the missteps, to know the exact points of failure (product features or price?). It may never be known unless the company is a learning organization, but if it was then maybe they would not have been susceptible to confirmation bias in the first place. More likely that product will just fade away without few answers as to "why," and the company will move on to the next big thing.

For McDonald's in 1996, their next big thing was the Arch Deluxe. Abandoning their core customer segment of kid-friendly and economically pleasing inexpensive meals for families, and we believe with limited marketing data, they launched a new burger. The new burger was not their regular fare but an upscale meal targeted at adults, and it came with more grown-up pricing too. Although no one would confuse McDonald's with fine dining, their foray into the segment of discerning customers might have been a surprise to management if it were not for the data from their focus groups. Focus groups are customers who get to preview products and provide feedback. For the Arch Deluxe, the focus groups were raving about the new menu item. So much so, that management spent an estimated $100 million on marketing, making Arch Deluxe one of the most expensive advertising campaigns in company history.[8] Gone was the fun-loving kid-friendly clown image of Ronald McDonald. Now Ronald McDonald was featured in ads doing more "adult" activities like shooting pool and playing golf. In retrospect, you can probably imagine why this did not work. However, at McDonald's, there was likely a ton of confirmation bias driving their decision and dismissing the poor sales data that might have changed the outcome. The product was canceled a couple of years later and McDonald's has had many other new products launches since, perhaps an indicator of lessons learned.[9] With such an expensive marketing campaign backing the Arch Deluxe and even millions more spent ahead of its launch, there could have been more than just confirmation bias driving their decisions. When a company, even a large one like McDonald's, spends $100 million or more, and there is hesitation to move forward, it can trigger another type of bias called the sunk cost fallacy.

Sunk Cost Fallacy

"Well, we have come this far, we might as well keep going."

Phrases like this seem straight out of just about any silly horror film, but they also haunt a lot of corporate boardrooms, stock investors, and gamblers. Earlier we noted most of us are susceptible to the notion of loss-aversion, that is, we hate to lose something more than we enjoy winning something, even of equal value.[10] In short, losing stinks no matter the amount and it hurts more than winning or just breaking even.

The sunk cost fallacy is embedded with this idea and usually arises when we have invested in something and we are at a decision point to either continue with that something or abandon it. The fallacy in our irrational logic is that because we have *already* invested in something, albeit money, time, and emotional energy, that our investment is somehow a factor in its evaluation. The investment we have made weighs as part of the decision to continue with the decision. This is incorrect. Just like our previous experience in flipping a coin onto heads does not impact the next toss—it is equally likely to be heads or tails—then we must not consider the previous investment already sunk to making a decision. It is a bit like saying "I will take my umbrella when I leave the house today, even though there is no forecast for rain because I already bought an umbrella." The decision to take the umbrella should be based objectively on the data—what is the probability of rain?—and not upon the fact that you have already spent money on an umbrella. A sunk cost is named so because the cost cannot be recovered, therefore it should not be a factor in a decision about the future. This is different from fixed costs, like the investment in a factory or equipment. Fixed costs can sometimes be recovered by selling the equipment or factory building. One strategy for avoiding the sunk cost fallacy is to ask yourself can I do anything to get that investment back?

When McDonald's was developing the Arch Deluxe, at any given point in the project they could have made the decision, based on the data, to continue investing in the product or halting its launch. The classic go-no-go decision is always imperiled when subjected to the sunk cost fallacy, and many times in our experience projects just do not have enough go-no-go checkpoints. There may be a few at the beginning, but once it is deemed a "go" at some point there is almost no turning back. The management team would have to consider the loss of millions of dollars of their investment. They would also have to confront previously making a bad decision.

The TV game show *Deal or No Deal* plays on this bias too. Contestants are periodically given a guaranteed offer—a cash prize. Taking the deal ends the game and their chance of winning a million dollars, but it provides them some amount of guaranteed payment. Continuing, however, means they could lose everything. The contestant's previous investment and belief that they could win it all clouds their objectivity about when to "no-go" on a decision. And more often than not, they should stop and take the certainty of the payment they are offered rather than going forward. In the five years of the show, there were only two million-dollar winners.

Hindsight Bias

"After seeing these results, we should have known better."

The truth is we have to rely on the best information we have at the time a decision needs to be made. We are better served to assess how well we processed that data (the data we had in the past) to the decision, rather than hypothesize how we could have made a better decision now. We could not have made a better decision in the past with information that was only available in the future unless of course you are Doc Brown and can time travel in a retrofitted DeLorean.

This is the essence of hindsight bias. It mixes the data we have now with the data we had in the past. If you find yourself with someone who is jumbling the data timeline like this, then this bias is in play. Further, if hindsight is impacting your ability to make a current decision, to question all the data you currently have because there is the fear of a repeated hindsight error, then you are experiencing a different type of data duping. The duper is using the data they do not have (it's in the future after all) to manipulate the decision you are making in the present. In our experience, these are stalling techniques where the duper is using the data to argue that "we do not know everything" or "we just need more data to make a decision." The critical question to ask in this situation is do we *need* to know everything to make an adequate decision. Do we have enough data to move forward, to limit our risks, and to optimize our opportunities? Note we mentioned to limit our risks since with almost all decisions there is always a bit of unknown risk. For example, your friends may not want to ride the newly installed roller coaster. They may say there is not enough experience from the operator to know if it is *completely* safe. But if you were a diligent data scientist you might have all the data about the design and testing of the roller coaster. You may have calculated the odds of an extreme event, like a critical bolt snapping in the middle of a ride, to know the probability of flying off into the parking lot after the third triple loop are infinitesimally small. For the person with the data and a rational approach to objectivity, they can make a fair and informed decision. Others may present the initial data (i.e., the risk of riding a newly installed coaster) and selectively ignore other data as a reason to nix the decision.

Group Think

"Well, if everyone else thinks it is a good idea, then . . . okay."

Sometimes it is hard to go against the flow when every-one else seems to want to make a different decision than you. But could this also happen when everyone has the same data? In 1951, a psychologist named Solomon Asch conducted his first of many "conformity" experiments. He wanted to understand the effects of peer pressure on an individual's ability to remain objective. It began when he presented a group with an illustration of three lines of different length and asked which one matched a fourth line. The task was simple and it was not a trick. One line of the three was clearly the same as the fourth.

Each participant had the same data and knew the true answer. How-ever, all of the group members except one—the true subject being tested—were insiders of the experiment and purposely choose another line as the correct example. How often did the subject change their mind and conform to the conclusion of the group? Remember the data are *very clear* about the correct answer, but astonishingly Asch found that 76 percent of the participants conformed to the group's consen-sus. Think about that for a moment. More than three-quarters of the participants were willing to go along with accepting bad data. The cor-rect answer could not have been more obvious. Why did they do that and what makes bad data more acceptable in a group situation? The good news is that this does not happen all the time and Asch further identified some characteristics of the group that made this more likely. These characteristics include the size of the majority holding the incor-rect view, the presence of at least one other dissenter, and how the conclusions were shared—either publicly or privately. Publicly shared responses gathered others in agreement and made conformity more likely. It may be uncommon for you to encounter such a perfect situ-ation. When the data are so clear you certainly will find others in the group will agree on what the data are showing. But what happens when the data are messy, with complex variations and murky conclusions? Does groupthink persist?

In the early 1970s, psychologist Janis Irving was helping his daughter Charlotte prepare a term paper for her high school history class. The topic was the Bay of Pigs, a failed invasion of Cuba by Cuban exiles who were secretly funded by the US Central Intelligence Agency. The homework assignment caused Irving to revisit Arthur Schlesinger's memoir *A Thousand Days*. In the memoir, Schlesinger wrote about his role as a member of President Kennedy's inner circle that planned the invasion over several meetings within the White House. Like Solomon Asch's subject, he also seemed to be the only member of the group who was using the data rationally and disagreed with plans for an aggressive invasion. What were the data? As Irving recounted in his book *Groupthink*, there were several miscalculations. One was the size of the invading forces, which totaled 1,500 soldiers. They would be easily defeated by the Cuban military, which included 25,000 members of the army and an estimated 200,000 militia and armed police. A few years earlier Cuba had just emerged from a revolutionary war. This was not just an army of overwhelming size, but also one that was battle-tested and ready to defend their newfound independence. With data like this, how could a group not see the obvious conclusion? Unlike Asch's setup, this was not a contrived experiment, but a real-life unfolding example of group conformity. Schlesinger wrote that "Our meetings took place in a curious atmosphere of assumed consensus." Individual members did not outwardly disagree with each other. Neither did they explain their reasoning for certain assumptions and recommendations. The data were not objectively presented and debated—a lesson for us all to observe. Schlesinger himself remained quiet in meetings even though separately he had written directly to the president about his concerns and disagreement.

The retelling of the events by Schlesinger had such an impact on Irving that he was unsure of how such irrational behavior could exist. He wrote "I began to wonder whether some kind of psychological contagion, similar to the social conformity phenomena observed in small groups [Asch], had interfered with their mental alertness."[11] Irving continued to study other groupthink-related fiascos and contributed heavily to understanding group conformity in the face of real data. For Schlesinger he recounted the episode as follows:

> In the months after the Bay of Pigs, I bitterly reproached myself for having kept so silent during those crucial discussions in the Cabinet Room, though my feelings of guilt were tempered by the knowledge that a course of objection would have accomplished little save to give me a name as a nuisance. I can only explain my failure to do more than raise a few timid

questions by reporting that one's impulse to blow the whistle on this nonsense was simply undone by the circumstances of the discussion.[12]

The lesson from Schlesinger's experience within groups making decisions is notable. Humans, as highly social beings, might overweigh the need for acceptance as an individual to combat the fear of disapproval from their associates. Overweighing acceptance instead of data can lead to criticizing the data, its sources, its methods of collection, and its very validity. This is because the data starts to compete with our personal relationships within the group. When data becomes the outcast, then data deception is just around the corner.

Selection Bias

"According to our survey of happy people, everyone seems to be happy."

In the fall of 1983, small start-up carrier Midway Airlines placed several ads in the *New York Times* touting an astonishing accomplishment. According to passenger surveys, "84 percent of Frequent Business Travelers to Chicago Preferred Midway Metrolink to American, United and TWA." It was a David and Goliath story given the small airline had been flying for only a few years, yet they had succeeded in winning the preference of 84 percent of travelers. The other miracle of this achievement was they had done this with only 8 percent of the flights between New York and Chicago. But wait. As the mathematical wheels in your mind are turning, and as by now you are becoming more astute at spotting data deception, you must be thinking that this does not add up. How can 84 percent of passengers prefer flying Midway Airlines but they were transporting only 8 percent of flights? Perhaps they preferred Midway but just could not book a ticket on Midway. Yes, that would fit the data. They really *wanted* to fly on Midway but could not, so they must have reluctantly traveled on less-preferred American or United. Imagine how American and United must have taken this news—there it was in plain-to-see data. Except for the fine print that accompanied the claim. The footnote at the bottom of the ad revealed the survey was not among *all* passengers who traveled between New York and Chicago. It was a subset. The survey was taken onboard Midway flights and therefore only included their passengers. Apparently, 16 percent of the passengers were rethinking their decision, but otherwise Midway found *most* of their passengers were happy to be onboard.

There are many other things to consider in surveys beyond who is chosen as participants and who responds. There is a whole discipline dedicated to good survey design. Without it, there can even be bias within the survey itself, such as leading questions, loaded questions, and the feared double-barrel questions, all of which can lead to data deception.

A leading question is perhaps the easiest to identify as bias since it usually has some qualifier, such as, "How well would you rate the response time of our world-class customer service center?" The qualifier is the term "world-class." What if the respondent does not think it was world-class? There is no way for that to show up in the data since the question's purpose is to rate the response time.

A loaded question usually makes some sort of assumption about the respondent and does not give them an option to change that assumption. For example, "How often will you recommend our product?" should include an answer that says "I do not recommend products." The question assumes the respondent recommends products and without this option, the data may be misleading.

The double-barrel question is when more than one issue is being asked within the same question. "Is our product value-priced and useful?" The answer could be one or the other, and maybe both, but really there should be two separate questions.

In addition to good survey design, there needs to be consideration of the survey sample, which is the group of participants to whom the survey is given. A random sample that is representative of the larger group is ideal. You do not need to look too far beyond the 2016 US presidential election to see the risks of not having a representative sample. Under-sampling certain groups can create blind spots in the data.

The data from surveys can be deceptive. As we have noted before, there are risks when averaging any set of numbers, and survey responses are no different. Most survey questions require respondents to choose a value on a scale. Typically, from one to seven but ranges can vary. Averages are easy to understand, however, surveyors often group bottom and top responses. In response to "On a scale of 1 to 7 how satisfied are you with your service?" a score of six or seven are reported as "Very Satisfied" when they might be different things.

Survey data can be very useful but it requires some good analysis too. It is easy to be deceived for the reasons above. An unbalanced survey pool for example that is not adjusted may provide you with a deceptive conclusion.

Surveys are not the only place where selection bias can be observed. Broadly, it speaks to the bias in how data are collected and curated. So-called "cherry-picking" data can bias analysis when only the data deemed best is selected and used.

Dunning–Kruger Effect

"The more I know, the more I know I don't know, but up until now I knew everything."

The Dunning–Kruger effect refers to how we might overweigh the amount of current knowledge (data) we possess and conclude it adequately represents all the data that is needed to make a decision *and* we discount the need for further data collection, analysis, and understanding. It is a shortsighted view of the problem and the data needed to solve it. When individuals are subject to the Dunning–Kruger effect, the current data can be used deceptively both by misleading the individual and by suppressing the consideration of other, perhaps more meaningful, data. A good test for this is when the person or group making a decision states they know the obvious answers although they have limited experience and knowledge. As they learn more, they find the obvious answers were not so obvious and realize the problem is more complex than they originally understood.

The antidote for this type of bias is a healthy dose of curiosity and humbleness among those seeking the solution to the problem. It is important to consider you or your potential data duper do not have all the information and may not even fully understand the problem. When the answer to a problem seems simple and obvious, then you may not have gone far enough to explore all the data.

Self-Serving Bias

"Everything we do is great. That other stuff is just bad luck."

Self-serving bias is the belief that what we do or control is responsible for all or nearly all of the positive outcomes and further, that the negative outcomes are not related to our direct actions. In other words, we make the good stuff happen and the bad has nothing to do with us. How does this show up in data and decision-making? Consider this example.

You run a retail services company and you are attempting to measure your customers' satisfaction and the factors that influence it. Where do you start? You might begin with your perspective of why your customers *should be* happy with your services. Things like your product, the pricing, and your ability to deliver consistently. Perhaps you bring a group of your company's managers together to join in on brainstorming the factors and they add . . . the product features, the benefits those features provide, and proudly how they (fail to) compare to the competition. Once you have a list of these factors you turn to your data analytics team and ask which factors are most important and therefore which ones should be focused on in the future to improve customer satisfaction. Do you see the mistake and bias that has been made? By first beginning with the things the managers control, the only data that will be used in the analysis are what the *managers believe* are important to satisfaction. The factors *they* control. What about other factors that might contribute or detract from satisfaction? Will they ever be measured in this scenario? Likely not, and self-serving bias creates this type of data deception by the *lack* of data not the misuse of data. It is the process of approach, which is framing the problem from an inward-looking self-imagined view that keeps the other data in the dark. The questions to ask here are what other factors might contribute to customer satisfaction are not within our control? And if there are factors that can be identified, how much do they influence satisfaction? Dealing with this bias requires looking beyond the data you have and looking a little further.

Survival Bias

"The ones that made it this far are the ones we study."

Survival bias is a special type of selection bias we noted earlier and most often involves time. As individuals, we are susceptible to this bias by way of our selective recall of past facts and our applying them to current situations. For example, let's say you want to determine the best method to assess a hockey player's skill. You might observe highly skilled players are those who play as a defenseman or center position.

Why? Because if you study the data of the NHL Hall of Fame, you know that most (54 percent) played this position. Oh, and it also helps if they are from Canada, of which 76 percent are among the inductees since 1980.[13] Is this a good way to assess talent and skills? Probably not, since many other factors determine skill and success on the ice, such as speed, agility, puck handling, and so on. But your vulnerability is to look only at the players who have "survived" long careers and were inducted into the Hall of Fame.

Survival bias of course goes beyond sports, often showing up in rankings and advertising of products' performance over time.

Belief Perseverance

"It seems the more facts we provide, the more you disagree."

Max Planck, the German physicist who is known as the originator of quantum theory in the early 1900s, wrote in his 1950 biography, "The new scientific truth does not triumph by convincing its opponents and making them see the light, but rather because its opponents eventually die and a new generation grows up that is familiar with it."[14] Perhaps Planck was struggling in the early days of physics. Struggling to learn how to convince others of the meaning of his data. His contemporaries included Albert Einstein and Walther Nernst, and they were known to hold long discussions about the newly advancing field and how to introduce their ideas to others. Perhaps Planck was among the early scientists to confront belief perseverance.

Belief perseverance, which is also known as the backfire effect, is the bias where individuals who have certain initial beliefs tend to dig in and fortify those beliefs, even when presented with contradictory data. The new information tends to strengthen rather than weaken their position. Psychologists vary in their explanations for this behavior. Some point to the cognitive reinforcements from learning information. The notion is that when we learn something and establish a truth, that little reward center in our brain fires and we feel good. In our mind, we guard this bit of knowledge because taking it away has the opposite effect. We have to concede we misunderstood something. We have to give in to a new truth and before we do that, we create a mental battle in our mind. We examine the new data. We undervalue it, and protect and overweigh the value of our original information. In other words, we have difficulty assessing new, contradictory information objectively. As the bias towards the original belief unfolds, we criticize the new data. We interrogate it. We try to discredit it. We give it more inspection and scrutiny than we ever gave to our original data. Even then, we may not accept it.

The problem with belief perseverance and data duping is the power of the backfire effect. The more data that is provided the more the truth is rejected. It is an irrational response and is the complete opposite of the more rational scientific process. Consider the example of heliocentrism first proposed in the third century BC by the Greek mathematician Aristarchus of Samos that the Sun rather than the Earth was the center of the solar system. This idea was not well known until Nicolaus Copernicus revived it in his 1543 publication *De Revolutionibus Orbium Coelestium* (*The Revolutions of the Celestial Spheres*). Even with a mathematical model and data to demonstrate the orbits of the planets, the idea was challenged by scholars, most notably Giovanni Tolosani on behalf of the church. Tolosani's response was typical of belief perseverance. He questioned the authenticity of the data and model characterizing it as an unproven theory. He attacked Copernicus's training and qualifications by stating he was deficient in his education in physics and logic.[15] Tolosani was protecting his previous beliefs and those of the church despite new data. Later the research on heliocentrism was continued by others, including Galileo Galilei, who, based on his observations of the moons of Jupiter in 1609, published his findings. What happened next? Similar to Copernicus, his work, his observation, and his data were challenged, so much that he was eventually summoned to the Vatican for an inquisition where he had to recant his findings and spend the remainder of his life under house arrest. The alternative for not recanting? The threat of being burned at the stake. Some data can create a profound response when others hold well-entrenched alternative beliefs. Eventually, in 1992 the Pontifical Academy of Sciences closed the dispute with Galileo some 350 years after his death in 1642. Data eventually endures the test of time.[16]

Illusionary Truth Effect

"We have heard this before; it must be true."

The illusionary truth effect is our tendency to believe information that we hear repeatedly. We are generally more susceptible to this effect when the information is plausible, but even when we hear information that we judge as less plausible, we are still a bit more likely to believe it.

How many times have you heard this? "Cracking your knuckles will cause arthritis" Or "Tide gets clothes cleaner than *any* soap" or "Staring at the sun will make you blind"?[17] Okay, that last one has some merit, but consider your initial reactions to the first two.[18] Do they sound familiar? Believable? True? And where did you first recall hearing these facts? Could you name a source or more importantly the data that backs them up? And now that you are reading them do you have doubts about their authenticity. If you do, you are not alone. Many statements like these are repeated every day and not necessarily in an intentionally data deceptive manner. But why do they persist? One answer lies in the marketing examples and the claims advertisers make. They repeat messages to build "brand-awareness" and to underscore the value of their product over another from a competitor. Does Tide really clean better than other brands? Maybe, it does seem plausible that Tide was more effective than other brands. To be fair Tide's competitors also made similar claims to win customers' beliefs, but with less memorable effect. It appears Tide's success in this 1950s-era campaign had more to do with the repetition than its plausibility.

The repetition of messages and information has been shown to increase belief in those statements. The number of times it is repeated is up for debate, but studies show that just one repetition can improve its illusionary truth.[19] An interesting finding of the research also suggests that repeating information can at times also shift belief perseverance, meaning it could help overcome some beliefs people had before they heard the repeated messages, even when those messages are not true. It seems the lessons from Marketing 101 that teach us to keep the messages simple and repeat them often hold true.

Joseph Goebbels was the minister of propaganda for Nazi Germany's Third Reich and he knew how to spread a message and at times misinformation. Consider this quote attributed to him: "If you tell a lie big enough and keep repeating it, people will eventually come to believe it."

The illusionary truth effect research suggests this is true to a certain extent. The limits are when the repeated information conflicts with other data the recipient already possesses and *validates* as true. You would be unlikely to convince someone that the Atlantic Ocean is the largest in the world or the Sun revolves around the Earth simply by repeating these false statements. Therefore, a big lie has difficulty being accepted, while a smaller more plausible one might succeed. The interesting thing about repetition and the illusionary truth is it requires inspection. Repeating something over and over may not always work, but sometimes it does. The Goebbels quote is no different. Goebbels was known for crafting all sorts of Nazi propaganda between 1933 and 1945 and it is believed he would have made the statement about repeating big lies. The problem is he didn't. Although this quote appears in thousands of online posts and other publications, research by University of Illinois professors Quentin J. Schultze and Randall L. Bytwerk could not find any credible source to show Goebbels actually made this statement.[20] The illusionary truth effect is alive and well.

The illusionary truth effect can contribute to our misunderstanding of the data we have for the same reasons we mentioned earlier. As humans, we have a need to bring order to disorder and understanding to misunderstanding, and sort meaning from chaos. The repetition of stories, and others' understanding of mysterious things, can embed themselves in our version of the truth. The dangers of swimming within one hour of eating, the ability to lose fat in targeted areas of our body, that volcanos erupt because the gods are displeased have at one time or another been repeated and accepted as true. As you are now likely coming to understand, the problem with the illusionary truth effect is not just how we process data, but how it can interfere with how we view *new* data and come to new understandings of the truth. When we do not use the new data objectively then we are vulnerable to being data duped.

Cognitive biases affect how you and others process data and they cloud our objectivity, especially when new contradictory information is presented. If you have been data duped, it may be connected to the bias in the data (due to poor data design, collection, or anomalies) but it also could be a result of our human biases. These biases create "data blind spots" preventing us from seeing new data and information. Many times, because these biases are cognitive shortcuts in our rational reasoning, we bypass more rational approaches, sometimes without even being aware of them.

AN OPEN MIND

Being open to new ideas and aware of our biases and those of others can be empowering. Knowing where we might have been misled and, perhaps, where we ourselves might be misleading others with data and our predetermined beliefs can be enlightening. This chapter is by no means an exhaustive look at how the human mind works, but rather a peek into the window of some of the mechanisms that propel our thinking. How we reason and process information, even under the best circumstances, with the most ideal data and good intentions, can still lead to data duping. Knowing more about how you got there will be valuable. As Socrates learned from the Oracle of Delphi, "a wise man is one who admits what he does not know."

KEY POINTS

- As humans we have an innate need for order versus disorder and as a result, we often seek to explain relationships between things that may be unrelated. No data required!
- Due to prior experience and learning, our mind can "fill in the blanks," or at least attempts to when there is missing information. We can do this visually such as with hidden images and misspelled words, and cognitively, when we prematurely jump to conclusions, sidestepping in our mind the need for validating data.
- Our processing of data can be influenced by previous beliefs (belief perseverance) and repetition of facts and untruths to the point that it persuades us from looking for validation (illusionary truth effect).
- The human mind has some shortcuts for processing data. We are capable of "quick decisions" and sometimes these appear hidden from our conscious mind as they occur.
- Several cognitive biases can contribute to data duping because we either process information incorrectly or fail to process the data at all. Biases that create blind spots in our rational thinking prevent us from seeing new data and at times may cause us to discard data. Becoming data duped can occur in the absence of data just as much as in the presence of true data.

9

Calling Out the Data Dupe

Addressing Misleading Data and Changing People's Beliefs

PEOPLE OR DATA

In previous chapters, we have explored bad data, misleading charts and graphs, the consequences of machine learning, artificial intelligence, and the pitfalls of ill-fit data models, and along the way we provided some tactical tools to avoid being data duped. What is next then? A common question we get from people is "How do you address a data duper?"

In "The Faces of Data Duped" section of chapter 1, we noted that sometimes the person presenting the information is not a deliberately devious data duper. They might be the unwitting data duper, a person who is misusing data without knowing it. Inexperienced with the pitfalls of data, they may not know how to question data and its conclusions.

The fallible data duper is the person who, with positive intentions, has constructed data conclusions through a combination of real facts, misleading opinions, and a sense of what *feels* believable. How many times has your uncle Dilbert claimed "the best corn comes from Iowa in June" or something similar because he has read an article that somewhat lauds Iowa farmers, or perhaps nostalgically, your uncle Dilbert grew up in Dyersville? Your uncle Dilbert has data but is not intentionally data duping.

The devious data duper may be more problematic to confront since often they are motivated to reach specific conclusions by selectively using data to support their point of view.

When you happen upon a data dupe you first need to determine if this is a result of people and their intentions or the data, which is blind from intentions yet can still mislead. Let's start with the data.

IT'S THE DATA, NOT YOU

Data can be messy, misshaped when plotted on a chart, and appear out of context when riddled with outliers. The first task of any good data scientist is to look at the data—really look at the data—to determine if it is usable. When it is usable, they must also decide how to rearrange it—fixing missing information and even transforming the data to account for wide ranges and apparent outliers. Not every data set requires this level of curation but when dealing with larger datasets and sampling data, it is a good practice. Data do not have a mind of their own, but to a data scientist some days, it may seem that way. When unattended, data can paint a different story from the truth. How different? Consider the US presidential election of 1936 and the *Literary Digest* poll.

In 1936, then-president Franklin Roosevelt was running against Republican challenger Alfred Landon, who was the governor of Kansas. *Literary Digest* was a popular opinion magazine and each year since 1916 it conducted a presidential election poll and successfully predicted the winner each time until 1936. Their process was simple, they sent polling postcards to each of their subscribers in addition to those chosen from car registration lists and telephone lists. They added up the votes for each candidate by state and allocated the electoral votes. The one with the most electoral votes of course was the predicted winner. It could not be simpler. In total, they sent out 10 million poll cards and received more than 2.1 million responses. This was a staggering amount given the US population was 128 million with a voting population of 45 million. The *Literary Digest* had completed a survey of roughly 1 in 20 voters.[1] In contrast, today's presidential polls use far fewer people in their samples; some contain only a few thousand. The *Literary Digest* had high credibility due to its previous polls and predictions. It is likely it felt little threat from the recently established Gallup Poll operated by the American Institute of Public Opinion. After all, the Gallup organization was only in operation since 1935 and managed to survey a mere 50,000 people for their 1936 presidential poll. They were barely a competitor, but that soon changed.

When the *Literary Digest* counted the votes, Landon was heavily favored with nearly 60 percent of the popular vote. Landon was expected to win 384 of the 550 available electoral college votes until he didn't. The real outcome was a landslide victory for Roosevelt, who won 60.8 percent of the popular vote and all but two states (Maine and Vermont), giving him 523 electoral votes, while Landon earned only 8. The prediction from Gallup? Gallup correctly predicted Roosevelt as the winner and thereafter launched Gallup into a polling brand that is still well respected.[2] The story of the 1936 election poll is a David versus Goliath tale. How could the *Literary Digest* with its millions of data points be so defeated by the smaller Gallup Poll? How did the data dupe them? Much like the folklore of David, it was about his precision rather than his size that mattered.

Recall the *Literary Digest* poll surveyed among their *current* subscribers, people who owned automobiles, and those who were paying for telephones. In 1936, the United States was still in the midst of the Great Depression. Unemployment had improved from its peak of 24.9 percent in 1933 but still was a miserable 16.9 percent in 1936. People were unemployed and poor. They had canceled discretionary items such as magazine subscriptions, stopped paying for a telephone (if they ever had one[3]), and many sold their automobiles. What that means is the people who responded to the *Literary Digest* poll were a biased population of mostly wealthy individuals. Those it turned out were aligned with and supported Landon for president. It was not a representative sample of the entire population, because it simply skipped a large cohort of people who welcomed Roosevelt's New Deal plan to revive the economy.

The Gallup poll in contrast was more representative of the population as a whole and provided better insight among all voters, even though their sample size was significantly smaller. The *Literary Digest* poll was larger but due to its biased respondents, it was not precise. Demographics had dramatically changed in a few years since the beginning of the Depression, which not only changed the group they were surveying but likely also changed the opinions of many people about how to lead the country out of its failing economy. The *Literary Digest* failed to see the flaws in sustaining their practice of polling—victims of "how we always do things"—and overconfidence in their large sample. The dupers were not people, but rather the data itself with its shifting numbers from previously reliable to horribly misleading.

When the data are the duper, the strategy to review their flaws are many of those we wrote about earlier in the data defense chapter. Of course, it is easy in hindsight to look at the *Literary Digest* poll and see how it could have been better. It is a good illustration of the need to examine the data before making conclusions. The data skeptics would have asked about the change in demographics. Sure, the poll worked in previous years, but why would it continue to work and in particular in 1936 with continued pressure on the economy? Perhaps there were other indicators the poll was not representative, such as analyzing how many responses came from each geographic region and comparing it to the country as a whole using other reference points such as the US Census data. To be sure the data were not deceptive, they needed to do an analysis and ask critical questions—how likely does this sample show the opinions of *all* voters and what would cause this result to be wrong? Knowing where the likely flaws are in data—revealed through analysis—can avoid a data dupe.

MEET PEOPLE WHERE THEY ARE

For every action, there is an equal and opposite reaction.

—Sir Isaac Newton (third law of motion)[4]

Sometimes the harder we work at convincing someone makes them dig in further with their opposing position. And the more we try to persuade someone the more they push back. Like Newton's Third Law, the amount of their resistance seems to reflect our effort to change their position. So, what is the solution? It is often best to avoid a direct confrontation of the data and rather first learn more about a person's interpretation and their intentions if any, in believing so. In this manner, you start with something you can both agree upon, which is their point of view, how they see the data supporting their position, and perhaps some insight into their motivation. Your uncle Dilbert believes this week's Mega Lotto is a good investment since several of his lucky numbers have not been drawn recently. His motivation is he wants to win; however, you know he has given little consideration to the math and probability. Similarly, someone buying a used car may be convinced the offered price is too high while the seller sees it as fair. Each has data to support their position—comparable car prices, costs for additional features such as leather seats and a top-notch stereo system—yet with the same data they have a different position and interpretation of the price.

In negotiating the price, each might argue they are looking at the numbers objectively—the data are the data—and without re-evaluation they may hold onto their respective positions and fail to agree, and the transaction will not occur. In the book *Getting to Yes: Negotiating Agreement without Giving In,* authors Roger Fisher, William Ury, and Bruce Patton write that it is inefficient to argue over each party's respective positions. It can lead to suboptimal outcomes and at times, without taking the time to separate the people from the problem, cause harm to ongoing relationships.[5] Rather, they suggest it is important to explore the interests of each side of a disagreement. What does each want? In the example of the used car, they may learn that each wants a fair price. With this newfound insight they can move the discussion away from their preset positions—the dollar amount the data are supporting—toward how each component of the data could be interpreted fairly. It may seem simple. The first step of meeting people where they are is to understand how they came to their current interpretation of the data and *why.* Agreeing first that you can appreciate and understand their interpretation of the data then opens up dialog and removes resistance. Let's look at another example that is subject to interpretation, sports.

One day your uncle Dilbert upon studying the baseball statistics, proclaims nostalgically that the baseball teams of 1998 and 1999 were the best of all time. And he has data to prove it. Are you being data duped? Maybe. Instead of pulling out your statistics and immediately taking the opposite position, start with meeting Uncle Dilbert where he is in his understanding and conclusion about "the best of all time." What numbers did he use to make this determination—what is the best way to measure the greatness of a baseball season? Total RBIs (runs batted in)? Or perhaps the similar statistic on total runs? Games won? If we are really looking at "greatness" maybe a better statistic is home runs, since this is the measure of the dominance of the batters over the pitchers. And on and on. Baseball is a sport with a lot of data and as a result, there are many ways to measure success. Uncle Dilbert has decided average runs per game is the best measure of baseball seasons and in fact, in 1999, the average runs per game for all teams was 5.14, a peak in the history of baseball from 1970 to the present, which averaged 4.45. Is there more to the data?

While there was a peak in those two years, a few other data points stand out. Although there was a dip in average runs per game in 2014 there is a trend of improvement through 2019 to 4.83 runs. Does this more recent rise indicate the current teams are heading towards greatness? Such observations could be used to challenge the claim of the 1998 and 1999 seasons. Further, is 5.14 runs much different than other years that were typically above 4?

Another way to look at the great seasons is by average home runs per game. In 1998 and 1999 it was 1.04 and 1.14, respectively. Since 1970 home runs have gradually been increasing and following a slight dip in 2014 of 0.86 had risen to 1.39 by 2019. At this point, you might be thinking differently and considering 2019 as a great season. You might also be convincing your uncle Dilbert, who maybe overweighted the 1998 and 1999 seasons due to bias. It was in those seasons when the news of baseball was all about home run record-setting.

In 1998 two standout players, Mark McGwire of the St. Louis Cardinals and Sammy Sosa of the Chicago Cubs, were chasing Roger Maris's 1961 record of 61 home runs. The data do support these seasons as standouts. Among the top 50 players the average season home runs were 35.56 and 36.82, respectively. What also stands out is the influence a few top players had on the overall averages. Removing McGwire and Sosa from the numbers shifts the average lower from 35.5 to 34. That may not seem like much, but when doing so those seasons begin to look more like the average and a new argument for 2019 as a great season continues to build. Can we conclude a great season is really so when it depends on a few players? What is notable is McGwire and Sosa were far above the numbers historically. Of course, they did set records and by a large margin—McGwire would hit 70 home runs, exceeding Maris's record by 9 or nearly 15 percent. Based purely on the data, Uncle Dilbert's conclusion has merit—these were great seasons. We have evaluated a few ways to look at the data and perhaps even Uncle Dilbert could appreciate he may not have been looking without bias at the performance of recent years. Just maybe the best baseball seasons are the ones ahead.[6]

Meeting people where they are is about avoiding ineffective confrontation. Often a data dupe occurs when there are large data sets beyond the math that one can do in their head, along with assumptions about data that they are free from anomalies and outliers. Also, there are times when an individual's bias, like your uncle Dilbert's nostalgia for a baseball season, can influence the interpretation. Knowing where a data duper is and where you are allows you to see the gap of under-standing and creates a path to have them step closer toward that data truth. A measured approach aimed at learning their assumptions and motivations along with deploying our earlier discussed data defenses (e.g., mean vs. median) can undo even the most determined deliberate data duper.

EMOTIONAL DATA

In the 1993 romantic comedy *Sleepless in Seattle*, the main character is distraught about whether she will ever find true love after a failed relationship. It feels like she is getting older and her prospects dim-mer. In a scene, a coworker says "It's easier to be killed by a terrorist than it is to find a husband over the age of 40." Annie, played by Meg Ryan, exclaims "That statistic is not true!" Annie's friend Becky (Rosie O'Donnell) in a show of compassion for her friend says dryly, "That's right—it's not true. But it *feels* true."

When data feels true, it is difficult to overcome no matter if you are a data believer or a data skeptic. Sometimes data seem so innate, so obvious that your internal senses—those that would cause you to ques-tion something—are suspended. The Earth is round, and the ocean's tides rise and fall due to the moon. These are true and also feel true, reinforced by our personal observations and repetitions over time. What about the notion you should not swim shortly after eating? Swallowing gum will stay in your stomach for years? And sharks feed at dusk, so kids, time to get out of the water! For many, those *feel* true too, and for the same reason. These statements can feel right because we can fit them into our observable data points that are often reinforced over time. Our mental heuristics cause us to fit the pieces into a probable puzzle and we mentally ignore the jagged edges of those few pieces that just do not always fit—like when a shark attack happens in the middle of the day. But what was feeding the emotional statistic about marriage after 40 that became embedded in 1990s culture and movies like *Sleepless in Seattle*? *Newsweek*.

In June of 1986, the front cover of *Newsweek* magazine showed an alarming headline and graph. "The Marriage Crunch: If You're a Single Woman Here Are Your Chances of Getting Married" it read, with a chilling chart showing your chances after the age of 40 were approaching zero. Yikes! The article inside titled "Too Late for Prince Charming?" was a demonstration of good data and bad math. Good data and bad math that still somehow feels right. The authors used US Census statistics out of context, carried them forward through time, and failed to adjust their findings for women who remarried and those choosing not to be legally married.

Here is a telling quote from the article:

> According to the [unpublished] report, white, college-educated women born in the mid-'50s who are still single at 30 have only a 20 percent chance of marrying. By the age of 35, the odds drop to 5 percent. Forty-year-olds are more likely to be killed by a terrorist: They have a minuscule 2.6 percent probability of tying the knot. —an excerpt from "Too Late for Prince Charming?" *Newsweek*, June 1986

Without data, the writers casually toss in "more likely to be killed by a terrorist." It made us wonder did they have statistics for unmarried educated women who were killed by terrorists or were they speaking in general terms of unmarried women? Of course not! Yet somehow, they wrote it with a mash-up of real data and conjecture, the latter getting credibility from the former. This, by the way, is a common technique among data dupers, leading people to consider if one fact is known and true then so is the other. Similarly, at least for the moment, many readers believed it could be true. And well it *felt* true. They knew at least one person older than age 40 who was not married, so maybe it was. Further, during the 1970s through the 1980s, terrorism, in particular airline hijackings, were in the news as a rising phenomenon. From 1968 to 1985, the number of worldwide airline hijackings was on average more than 20 each year. The number in the United States was fewer, although that did not matter to the concerned readers of *Newsweek*. The total number of terrorist deaths from all causes in 1986 when the article was published was only one. Yes, only one! How the authors compared their (faulty) marriage statistic of 2.6 percent to one total fatality and concluded marriage was less likely is an amazing case of data duped. Although it was mostly the confirmation bias of *Newsweek*'s readers bolstered by over-reporting of recent events—both terrorism and the growing number of women in the workforce—that allowed this data dupe to persist.

Twenty years after the original article, *Newsweek* published a retraction partly titled "Why We Were Wrong in the Article *Rethinking the Marriage Crunch*." We are not the first to point out the data deception of the first article. It has repeatedly been refuted. In 2000, *Discover Magazine* called the article one of the 20 blunders in science—the list also included Chernobyl and laboratory-created Killer Bees. It has been referenced several times in pop culture, in articles and in movies as a reference point for decisions people make and their consequences. Often those decisions include data. We *feel* like we are using data but more often we are choosing our confirmations of our fears and going with our emotions. As the *New York Times* put it, "For a lot of women, the retraction doesn't matter. The article seems to have lodged itself permanently in the national psyche."[7] Data that feels right can be like that.

LOGICAL DATA

Winning over someone's logical point of view may seem, well, logical. This is best applied to scientific disputes. Not just experiments performed in laboratories, but anything that can be framed using the scientific method, any question where a hypothesis—a likely conclusion—can be stated and tested through observation and the collection of data. Do Millennials play the video game *Fortnight* more than Generation Z? Do they spend more money online and are they a viable marketing segment for "gamified" marketing campaigns? Are more babies born in the summer months? Are tornadoes more likely in the month of April? And so on.

Logical conclusions based on data do have some limitations. The outcome of collecting data to confirm or deny a hypothesis can result in "Yes—it is true," "No—it is not true," and many times "Maybe—the result are inconclusive and further analysis is required." Using the scientific method may not end up with the results *you* intended, however, when confronting another person's point of view, you could at least agree on the results of the study. The challenge, of course, is it requires the person or group you are opposing to understand and accept the scientific method and the design of your analysis. Further, they, like you, have to be willing to accept the results as an objective arbiter of whatever point of view is disputed.

If you are confronting a data duper and they have agreed to accept logic, then good! You are off to a great start. However, sometimes the person with the opposing view wants to use logic in the form of a debate. In this case, we are still optimistic about the approach, provided there is a prevailing commitment to logic.

In a debate, a common structure for one person to take is to state a position in three parts—a major premise, a minor premise, and a conclusion. A data version of this structure might look like this:

(Major premise) Outliers can influence the value of the mean.
(Minor premise) The data set we are reviewing has several large outliers.
(Conclusion) The mean is not a good representation of the typical values of these data and should not be used without adjustment to make statements about the overall population of interest.

This structure is a sound way to arrive at a conclusion, or inspect a data duper's approach to their conclusion. Did they have a premise for their conclusion and if so, what was that premise. Of course, during this inspection, you may discover there is a problem with the logic being used. The premise may be flawed.

The movie *The Great Debaters* tells the story of a newly formed college debate team in the 1930s. This is a fantastic movie not only for its central plot but also for some of the lessons it teaches about how debates are won. Most characters in the movie do not use data directly, however, the technique is the same—present a logical premise supported by facts (data) and lead your opponent to a sound conclusion. On the other side of the debate is how to unwind any faulty logic (premise statements) and dislodge the conclusion for not being based on facts. Professor Tolson, portrayed by Denzel Washington, demonstrates this in a scene when one of the students is debating the New Deal and why it is necessary. The professor states:

You gave a faulty premise so your . . . logic fell apart.
Major premise—the unemployed are starving.
Minor premise—Mr. Burgess [your fellow student] is unemployed.
Conclusion—Mr. Burgess is starving.
Your major premise was based on a faulty assumption.

—Professor Tolson in *The Great Debaters*, 2007

Obvious to the student at this point in the scene is that her classmate is unemployed (a student) and clearly not starving. The premise statement that *all* unemployed people were starving was wrong and without a factual premise. The conclusion is also incorrect. Students who have participated in debate class will know this as the classic logic fallacy and it is a good tool to unwind spurious data conclusions presented by data dupers.

Beyond the stories told in movies there are other examples. Breakfast is the most important meal of the day! If that sounds familiar you are not alone. In the early 1900s, John Harvey Kellogg, a physician who advocated for better nutrition in diets, is often credited with originating the phrase. He is also the brother of Will Keith Kellogg, known to his friends as "W. K." and known to the rest of us as the founder of W. K. Kellogg company and inventor of cornflakes breakfast cereals and many others. The phrase has become synonymous with other marketing campaigns and has been ingrained in our thinking about both the importance of breakfast and healthy living. The longevity of this claim may also have contributed to several observational studies about the effect of a regular breakfast on weight loss. In a 2002 study, researchers showed a correlation between people who eat breakfast every day and sustained weight loss.[8] The study has been misquoted many times with the conclusion that breakfast can contribute to weight loss—even though it requires eating more food than skipping breakfast.[9] The authors did note their work was not a prospective study and did not control for other weight loss factors, such as an increase in activity among those who regularly ate breakfast. However, the simplicity of the solution—eat more, eat regularly, and lose weight—seemed to overpower later readers into accepting the false logic of the research. The logic followed:

Premise—Breakfast cereal is a healthy meal.
Premise—People who are not overweight eat breakfast.
Conclusion—Eating breakfast will help you lose weight.

Structured like a logical debate in this form it is clear how the premise fails to be factual. Yet of 72 subsequent studies that cited the 2002 research, more than half overstated the association between breakfast and weight loss.[10]

We have demonstrated how logic can work for or against you and the risks of a non-data-based premise. Approaching a data duper with logic, the inquiry might include: What data do they have to support the conclusion? Is the data from a credible or biased source? Are there defects in the data and how it was prepared (e.g., were outliers considered and removed?). What premise(s) were used to understand the data? How does the data support the conclusion? What if the data were slightly incorrect, would they arrive and support the same conclusion? Finally, was the conclusion and the data a coincidence, an indirect correlation, or (hopefully) a result of good science?

Using logic to unravel a data duper can work and it also can have pitfalls. A memorable way to keep this in mind is to consider the illustration from the comedy troupe Monty Python in the movie *The Holy Grail*. In a mockery of both the Salem witch trials and a host of other poor decisions throughout history, the segment shows how foolish people can be when they make decisions seemingly based on fact and logic.

The scene opens with a marauding crowd dragging a perceived witch in front of the noble knight demanding the witch be burned at the stake. The crowd shouts their evidence about how she *looks* like a witch and therefore must be a witch. This is a good example of confirmation bias. However, the knight convinces the crowd there are logical ways of telling if someone is a witch and explains it as follows:

Premise—Witches burn and wood also burns.
Premise—Wood floats in water and ducks float in water.
Conclusion—If a woman weighs the same as a duck, then she is a witch.

The mockery is complete as they place the woman on a balance scale with a duck and she weighs exactly the same! Perhaps there was a flaw in their data collection and the measurement needed to be repeated? The scene has all the elements of data duping—an assumption ahead of observation of the facts, use of logic with a flawed premise, and error in the collection of data. As the two noble knights congratulate themselves on their cleverness, they remark, "Who are you, who is so wise in the way of science?" It is classic comedy. They are convinced because they are following logic and not responding emotionally like the townspeople, they have arrived at the right conclusion. They could not be more wrong.

In the real Salem witch trials that began in 1692 in colonial Massachusetts, hundreds of people were accused of witchcraft. The trials adhered to accepted English law and followed a logical process. There were accu-sations, witnesses, testimony, and trials. The flaws were in the testimony, and data collection, of the hysterical townspeople claiming to be hexed and physically hurt by the "spells" of the witches. The hysteria was a result of confirmation bias. Few could see beyond their pre-judgments, emotions, and flawed logical arguments. Sadly, 19 people were found guilty and executed.[11] The conclusion arrived before the proper data.

IN THE END . . .

The best way to change someone's opinion about something is to make sure they don't have an opinion in the first place.

—Anonymous

This book is about how best to understand data—how to avoid being hoodwinked by misleading information that contributes to ineffective decisions. When working with people in the context of a decision, you find yourself feeling more like you are negotiating with them to see a certain view of the data, or alternately they are pursuing you to see their perspective. You might think of data as a plain and non-fungible matter—there is no substitution, no interpretation. However, as we noted earlier, data can be transformed and manipulated to support a particular position. Outliers, for example, may conveniently be included to bolster a number, when in reality they are not representative of the typical observation. Contrast the current 20-year stock market return (S&P 500) of 5.9 percent to the last 5-year average of 15.9 percent—a few good recent years can dramatically swing the numbers.

There are also situations where the data are difficult to interpret: large data sets—more math than you can do in your head; a data set with large outliers; a skew that tosses the average one way or the other. Then when interpreted by novices, some with motivations for a particular conclusion—wham! Data duped. Weight loss? Some miracle drug? A miracle sports team? A miracle year? The year of all years? The worst year for hurricanes? These are all examples of difficult things to quan-tify. Take the hurricane example further. How is the worst hurricane year measured? Wind speeds, number of storms, total rainfall, flood-ing, total damages in dollars (adjusted for inflation we hope), deaths, disruptions, lightning strikes, and number of households within fifty miles of the eye of each storm? Five hundred miles?[12] Defining how to use the data—agreeing on the terms of worst hurricane year or the best season in baseball—are the first steps in calling out a data dupe. Knowing *what and how* something will be measured helps to build or defeat the way numbers are used.

Another example might be investing in your uncle Dilbert's coffee shop—he says it's a sure thing and he has the *data to prove it*. Proforma financial statements on a spreadsheet are a good target for being data duped. What were the underlying assumptions? How did they arrive at those numbers? Do the numbers account for uncertainty—people, prices, product, promotion, and so on? What is the very best to be expected? How about the worst? Can there be an equivalent of a 100-year flood in a business plan? Err . . . a 100-year pandemic that upends even the best business ideas? If so, what are three scenarios that could occur—best, worst, and most likely. Investing in a business is a risk; as the saying goes if it were that easy everyone would do it. The question (and again this is a book of questions) is the risk known and reasonable? Does the probability of the expected return account for the measured risk? How does that risk compare to other investments—a US Treasury Bill, a corporate stock, or a junk bond? If you have a reference point, then you have perspective. And having perspective is everything. You are no longer data duped—you are data-informed! With a mind wide open, you can better assess the risks and the rewards even when the numbers are complicated.

We have discussed approaches to data dupers who are emotionally connected to their facts and others who fail to see the flaws in their logic. The bandwagon effect may have already taken over, causing people to see the popular conclusions over the logical, data-supported ones. The bandwagon effect may also contribute to people overly defending a position that is not based on data. The strategy of meeting people where they are will help you explore which of these attributes are contributing to the data dupe. Using your newfound data defense from earlier chapters as tools to better evaluate the numbers will help in changing factual understanding, even when it might be your own. In the end, it may be difficult to change a person's point of view even with the best logic and best data. Sometimes the best way to shape someone's point of view is to start before they have one. Afterward, it may just not be possible to change.

KEY POINTS

- Keep an open mind about data dupers. Some are deliberate dupers, while others are unknowingly being duped by the data itself.
- Meet people where they are by learning how they arrived at their conclusions and interpretation of data. Learn about their motivations before exploring the data.
- Defeating data dupers is similar to debates and both can include flawed emotional and logical missteps.
- At times changing a person's point of view needs to begin before they settle into a position. Once they have confirmation bias, the bandwagon effect and the willingness to prove their conclusion may make the task impossible.

Put a Little Data in Your Life

All truths are easy to understand once they are discovered; the point
is to discover them.

—Galileo Galilei

ENLIGHTENMENT

What would science be without discovery? What would discovery be
without curiosity and skepticism?

There are periods in history when there was little advancement in
science and knowledge in general. The period following the fall of the
western Roman Empire for example is referred to as the Dark Ages for
its decline in advancing knowledge. There are other points in history
when pivotal inventions cascaded to further develop other ideas, such
as Edison's sustainable electricity (1880), the steam engine (1781), the
transistor (1947), and the internet (1983).[1] Innovation is not a linear
process over time, rather it has spikes of great ideas and then ebbs of
slower progress. The analytics revolution is following a similar path. As
we noted, the convergence of inexpensive and faster computing power,
improved analytic techniques, and the tremendous amount of data are
fueling changes in nearly everything we do. A world more immersed
in data requires you to be more prepared, to be skeptical, yet curious,
wise, and enlightened.

The Enlightenment was a period in history between the mid-
seventeenth and the eighteenth century. It was a time of historic

195

transformation and was considered a cultural and political turning point. Many of the ideas that arose in this period—individual liberty, political and religious freedoms, and challenges to institutions of authority, namely the church and monarchies—eventually contributed to provoking revolutions across Europe and the United States. The uprising of people was an uprising of ideas, of awareness and knowledge. The period had impacts on a broad range of disciplines from art to politics to philosophy. All good things, but, of course, it was the science that was most interesting.

During the period of the Enlightenment, there were several important scientific discoveries such as Kepler's planetary motions (1619), Newton's laws of motion (1687), and the discovery of magnetism and early forms of electricity and batteries by Leyden (1745) and Franklin (1751). Also, during this period the populous (commoners) were becoming more literate and interested in learning. It was vogue to be knowledgeable and to be less reliant on institutions that directed people's ideas, and at times limited their access to knowledge. The influential leaders of the Enlightenment, known as the *philosophes*, sought to fill the need of the people by spreading knowledge to a greater number of readers. One such accomplishment was the encyclopedia.

Beginning in 1751 and continuing until 1772, a group of contributors led by France's Denis Diderot published the "knowledge of everything known" in one set of work that continued over 17 volumes of text and 11 volumes of plates. In total an impressive 74,000 articles. Diderot commented on the encyclopedia as seeking to change the way people think as he wrote "The purpose of an Encyclopedia is to collect the knowledge scattered over the surface of the earth, to expose the general system to the men with whom we live, and to transmit it to the men who will come after us . . . and for common people to inform themselves and know things."

However, the most important contribution to the Enlightenment might have been by Englishman Sir Francis Bacon. As an early contributor, he introduced a new approach to learning in his 1620 work called *Novum Organum*, translated from Latin as New Instrument or New Method. In this work, he proposed that knowledge is gathered from observation and testing to seek the truth, rather than the prevailing approach, which was to start with a truthful premise and then work towards a conclusion. It might sound like the same idea but was notably different. It changed how people arrived at truthful conclusions—rather than starting with a premise that the Sun revolved around the Earth, it started with the observation that the Sun and the Earth were in motion. Further observation would lead to the correct conclusion—the Earth revolves around the Sun. It was the data finding the truth instead of the truth trying to be disproved by the data. The latter is flawed with the assumption that something is true until data prove otherwise. It was a significant change in how people sought knowledge and likely contributed to the exploration of ideas during this period in history. Rather than seeking knowledge and truths from institutions of authority (Church and State), the Baconian Method became the framework of the scientific process—the process of developing a hypothesis, subjecting it to observation and testing, and then arriving at a factful conclusion.

The Baconian Method became so accepted by the philosophers and scholars of the time that they thought any other means of conclusion was incorrect. At that time to accept something as true without evidence, obvious as we see this belief as wrong today, was common. As a result, the Enlightenment is often referred to as the Age of Reason. And some add the age of *right* reason, since, like the examples of errors of false premise during debates, the process requires good facts, along with its disciplined process. To characterize the significance of Bacon and later work added by Immanuel Kant, historian Thomas Osborne wrote:

> [The Enlightenment] in its broadest, most banal, sense, the notion refers to the application of reason to human affairs; enlightenment would be the process through which reason was to be applied to all aspects of human existence, above all in the name of freedom . . . The great thinkers of the Enlightenment all believed that reason as opposed to superstition or dogma was the one sure basis of a free and just society.

—Thomas Osborne, 1998[2]

As we experience the unfolding of the analytics revolution and the spilling of data into every corner of our lives, we often accept fact without question, reason without doubt, and truth without skepticism. As we do, news headlines, marketing advertisements, and social media posts have a power similar to medieval authoritarians. The philosophes of the Enlightenment and in particular Bacon's approach to finding knowledge is a similar parallel to our modern times and an illustration to avoid being data duped. The ideas of the Enlightenment created a culture that encouraged questions, sometimes to the dismay of others, and shepherded great discovery guided by curiosity. The lessons of *Data Duped* we hope prompt a similar curiosity driven by statistics and an understanding of data science and empowers you, too, as the Age of Enlightenment empowered Western civilization.

DELIBERATE DATA DUPERS

We would be remiss not to include deliberate data dupers in our closing chapter. In general, we have focused more positively on the data dupe as mostly an outcome of a series of unintended consequences—poor data, faulty assumptions, bias, the problems of both small and extremely large data sets, and the shortcomings in the skills of those trying to make sense of it all. However, at times we have to remind ourselves as data professionals that not everyone views the world with as much data as we see. We remind ourselves that many things are easy to accept as true even with the minimum of supporting data. There is a vast amount of data out there and do we really need to look at data for *everything*, when the answers seem obvious? Sometimes yes, especially when those "obvious" answers are motivated by someone who benefits from getting someone to that conclusion without all the bother of the pesky task of looking at the data. Good examples include the stereotypical used car salesman who would rather rush you to sign the "deal of the day" instead of having your mechanic take a few measurements, or the one-sided late-night infomercials. Thus, we also need to remind ourselves there is a lot of data duping that occurs with little data at all, and often this is done by the deliberate data duper.

We introduced the deliberate data duper in chapter 1 and have not revisited them much since. The reason is the deliberate data duper is often doing their nefarious dark art of data deception almost magically—without data. Without data, there is not much for us to comment on or to show how they are using the numbers incorrectly. They are simply not using numbers! And if there is one thing you have learned so far it is that conclusions *appearing* to be based in fact but failing to provide the data are a clear sign of a data dupe. It is the snake oil salesman promising the cures of the world with nary a mention of the facts, research, or data to support it.[3] And for this, we could not give much attention, other than to make you more aware and skeptical. Our research has found this type of deliberate data duper in all the likely places and targeting our weakest vulnerabilities. Consider late-night television ads and weight-loss programs or alternative medicine focused on the benefits of *natural* health solutions such as vitamin supplements that are often no better than the nutrients we get from our everyday diet. Add the get-rich-quick schemes that also seem to be a sure thing, with little risk, and you have a full complement of deception.

At times we have also observed a type of deliberate data duper who will use data selectively. Inclusions and exclusions of data, especially outliers as we have often mentioned, can have big impacts on averages. But other times, the notion of credibility can be built when a few reference numbers are included. An example is a short-lived misinformation piece about the effect of vaccines on women's infertility that demonstrated how a data dupe sometimes starts with a grain of truth.[4] The story was introduced online and it noted a true fact that some women experience heavy menstrual periods following vaccination. The information was anecdotal yet plausible that it could be connected to vaccinations. Although this was unlikely to have any effect on infertility since heavier menstrual periods do not equate to infertility, the story grew further online. It was the result of a false association—a correlation with little observation and a scant sample of just a few people. Have you ever had a similar experience? You note something unusual happened and immediately try to think what might have caused it. The neighbor's house gets struck by lightning . . . and well they did just get a new car, maybe it's related. It is this type of thinking that starts a good viral story.

While your neighbor's misfortune gets chalked up to bad luck, the story of vaccination and infertility continued to grow. It got boosted by an influencer on social media and a further boost when a traditional news channel repeated the story. Perhaps the news channel intended to draw attention to a spurious claim, however, there are many examples where even the negative reporting of a false viral idea helps increase a story's reach to a broader audience. After a few weeks, the story faded away, but it demonstrates how easily a deliberate data dupe can occur, and in the case of health care–related dupes, we wonder how many people were harmed. Although many social media platforms make an effort to fact-check and filter misleading information, examples like these can be difficult since they blend truth with opinion. When a post factually puts two data points side by side, such as "29 people in nursing homes died following vaccinations," the post is guilty of a fallacy of logic. They are misleadingly relating two events, while not explicitly stating the connection. The use of data makes the dupe even more believable.

It is also worth noting that a growing participant in the deliberate data duper category are governments. We wrote earlier of Finland's initiatives to teach critical thinking beginning with their elementary school children as a result of foreign misinformation campaigns. Following the US 2016 presidential election, the topic of foreign influence in politics became widespread, and we believe improved how many people judge political advertisements and other politically related information. Fact-checking these topics by major news media increased and niche players such as PolitiFact.com and Factchecker.org focused on this type of misinformation. However, what Finland discovered as early as 2014 was misinformation campaigns were targeting a broad range of topics, not just who should be voted in at the next election. The purpose it seemed was to spur division among groups of people and promote falsehoods on sensitive topics. We see these purposes continuing into recent topical debates such as the effectiveness of vaccines and the perceived dangers of 5G mobile phone networks. Unlike the profit-motivated 2:00 a.m. infomercial selling super vitamins to (falsely) protect you from cancer-causing free-radicals, foreign actors also have something to gain from stirring divisive topics from budget investments in national defense to France banning work emails after 6:00 p.m. Defending against a data dupe, as Finland has begun, has become an imperative in education on par with reading and math.

EVERYTHING ELSE

Why did we skip certain things? This is not a book about debunking myths. Although fun and entertaining we wanted to be purposeful about how we guided readers more towards how to *ask* questions rather than providing answers. In our observation, many myths lack data in the first place, yet it still surprises us how often things can be accepted as truthful without supporting data. If there is nothing else you take away from this book, it should be this—there are data for *everything*. Someone who says there are no data might already be down the path of a data dupe.

We want you to be data-empowered and hopefully more data-aware from the material we have presented. We hope you are more knowledgeable than before and have a greater understanding of data and statistics and all the ways those two things combined can mislead. We have shown examples that unfortunately are patterns that will repeat—in the news, in advertising, at work, and in your everyday encounters with your friends and family. It is easy to misjudge numbers and for others to do the same. We have our bias, and in a similar way, so do some of our data.

Throughout this book, we have made several historical references and hopefully, you have noticed the purpose. Certainly, there have been many examples of data blunders in the past—Napoleon's march to Russia,[5] Boeing's 787 Dreamliner lithium battery issues,[6] and the demise of the Mars Climate Orbiter.[7] Each of these relied on data in a way that led to confidence in decisions and unfortunately to terrible outcomes. We have used history to present memorable examples and to demonstrate that people, sometimes very smart and intelligent people, can be hoodwinked by data. They may be either misguided by their (mis) understanding of the data or drawn to false conclusions by their bias like the sailors lured into the rocks by the songs of the Sirens. History shows how it is possible. For some, it may leave them with overconfidence that history teaches us a lesson, and once learned we, therefore, have protection from that particular type of data duping. There is some truth in this sense of confidence, yet we still find it hard to believe nearly every year we continue to see data used to create deception from news headlines, marketing ploys, and specifically Ponzi schemes of even greater size and sophistication. Recall a Ponzi scheme is a fraudulent investment that lures people by paying often extraordinary returns on their money that is taken from other investors. Ponzi schemes continue deceiving individuals as a group by an average of more than 1 billion dollars each year. This occurs even after the highly reported Madoff Ponzi scheme in 2008. History such as recounting how Madoff stole millions of dollars can serve as a guide although these repeated schemes remind us history is not enough to change future decisions of some people. This is why it is important to explain past events where data are used as a deceptive tool. (See figure 10.1.)

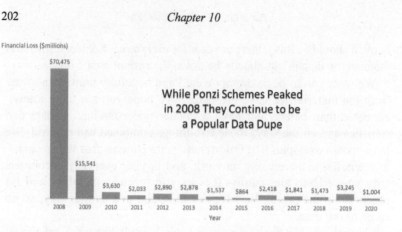

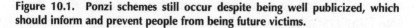

Figure 10.1. Ponzi schemes still occur despite being well publicized, which should inform and prevent people from being future victims.

PUT A LITTLE (MORE) DATA IN YOUR LIFE

We hope you might experience life a little bit differently following the reading of this book. Perhaps a growing sense that there really *are* data all around you. When you go to the grocery store and wonder why the milk is located in the back. (It is not random.) Is there a better, more efficient way to load passengers onto airplanes? (There is.) How can an auto dealer sell a car with zero percent interest when the lowest bank rates are higher? (They cannot.) All of these are variations on being data duped. The grocery stores are using low-priced goods positioned in the store in a way that yields more sales from other products bought while walking farther into the store. The airlines have several ways to optimally load passengers onto airplanes but would miss out on revenue they gain from selling premium-priced priority boarding passes, while also selling discounted tickets.[8] The auto dealer simply is increasing the price of the car to decrease the amortized interest costs and using the numbers to grab your attention. With a little more data in *your* life, you might recognize these patterns of data deception and, more importantly, be enabled to make more informed decisions, knowing often the data are correct, the problem is how they are interpreted . . . and by whom.

The future is bright and full of data shaping nearly everything we do. We believe the analytics revolution will bring positive change in many industries from manufacturing to health care and beyond. The growth of data continues to expand at an ever-increasing pace, with some companies still exploring how to use it all. By 2025, Seagate, the maker of data storage devices in collaboration with IDC, estimates the world's data, what they describe as the Global Datasphere, will grow from 33 zettabytes in 2018 to 175 zettabytes. It is not clear how they made those estimates, but we would agree at a minimum data are growing at an exponential rate. This of course is driven by several factors, including a more digitally connected world. Consider these data points:

- 5.11 billion unique mobile users, among 7.6 billion people
- 4.38 billion internet users
- 3.4 billion social media users
- IoT (internet of things connected devices) from smart watches to coffee makers is expected to increase to 43 billion by 2023, a 300 percent increase from 2018.[9]
- Businesses will continue to evolve to more data-centric offerings and decision-making since currently only 31 percent surveyed consider themselves data-driven.
- 50 percent of companies are using at least one AI/machine learning technique in their product offerings or operations. Although mostly in technology and telecom, the cascading impacts can influence our retail experiences and most customizations driven by Amazon-like recommendation engines. Machine learning has yet to reach high adoption rates and its untapped potential.
- 38 percent of jobs are estimated to transform or transition to automation by 2030.[10]
- Cloud computing, an indicator of both a transition of business models and the deployment of AI and machine learning, grew 34 percent between 2019 and 2020.[11]
- Recommendation engines are becoming more common. Online retailers can change and customize the products you view *and* their prices every 10 minutes.[12]

On the topic of recommendation engines, are we that far away from real-time price changes in the grocery store? Will ice cream be more expensive on warmer days or will the prices decline for peanut butter and jelly when the trends show it declining on non-school days? And is this real-time pricing an optimal customer experience or a data-driven ploy at deception of the *real price*? Although it may not be too different from the slower process stores use today to seasonally adjust prices and manually replace price tags. What is better? What is fair? These are types of questions we hope *Data Duped* helps you understand and expect from the future of data.

***

A few parting thoughts. When in doubt, ask the basic questions. Verify news and spurious claims by at least one other source—better if there is more. Question bias, yours and others'. Question motivation, yours and others'. Be a data steward and a data skeptic. Be a champion of the truth. Be curious, especially when *the data sound too good to be true*.

During the Enlightenment, Rene Descartes coined the phrase "I think therefore I am." We would add "without data therefore I am not." Data exist in all we do. That is inescapable, and, as we have repeated many times, it is something that will continue to grow and drive our world. Becoming "data enlightened" can be empowering, and if you remember one lesson here it should be this:

When something seems too good to be true, the claim too real to be believed, when the numbers evade the truth, then you might have been *Data Duped*.

Notes

CHAPTER 1

1. Åse Dragland, "Big Data, for Better or Worse: 90% of World's Data Generated over Last Two Years," *Science Daily*, May 22, 2013, https://www.sciencedaily.com/releases/2013/05/130522085217.htm.

2. Robert J. Moore, "Eric Schmidt's '5 Exabytes' Quote Is a Load of Crap," RJMetrics.com, February 7, 2011, https://blog.rjmetrics.com/2011/02/07/eric-schmidts-5-exabytes-quote-is-a-load-of-crap.

CHAPTER 2

1. "Evolution of the Mainframe," IBM, accessed September 22, 2022, https://www.ibm.com/ibm/history/exhibits/mainframe/mainframe_intro.html.

2. Frank da Cruz, "The IBM 1401," Columbia University, accessed September 22, 2022, http://www.columbia.edu/cu/computinghistory/1401.html.

3. Jeff Nilsson, "What the Operators Overheard in 1907," *Saturday Evening Post*, June 30, 2012, https://www.saturdayeveningpost.com/2012/06/operators-heard-1907.

4. Matt Day, Giles Turner, and Natalia Drozdiak, "Amazon Workers Are Listening to What You Tell Alexa," *Bloomberg*, April 10, 2019, https://www.bloomberg.com/news/articles/2019-04-10/is-anyone-listening-to-you-on-alexa-a-global-team-reviews-audio.

5. Cambridge Analytica created an app downloaded by 300,000 Facebook users that, among other things, gave Cambridge Analytica access to not only the profiles of those individuals but also all of their other Facebook friends. This allowed them to create a massive database claiming to profile 50 million Facebook users.

6. Elizabeth Aguirre, Dominik Mahr, Dhruv Grewal, Ko de Ruyter, and Martin Wetzels, "Unraveling the Personalization Paradox: The Effect of Information Collection and Trust-Building Strategies on Online Advertisement Effectiveness," *Journal of Retailing* 91, issue 1 (2015):34–49, ISSN 0022-4359, https://doi.org/10.1016/j.jretai.2014.09.005.

CHAPTER 3

1. Fletcher Schoen and Christopher J. Lamb, "Deception, Disinformation, and Strategic Communications: How One Interagency Group Made a Major Difference," *Strategic Perspectives* 11, Institute for National Strategic Studies (INSS). National Defense University Press, Washington, DC, June 2012, https://ndupress.ndu.edu/Portals/68/Documents/stratperspective/inss/Strategic -Perspectives-11.pdf.

2. Ibid

3. John P. A. Ioannidis, "Why Most Published Research Findings Are False," *PLoS Medicine* 2, issue 8 (August 2005):696–701, https://doi.org/10.1371/journal.pmed.1004085.

4. John LaRosa, "Top 9 Things to Know about the Weight Loss Industry," MarketResearch.com, March 6, 2019, accessed September 24, 2022, https://blog.marketresearch.com/u.s.-weight-loss-industry-grows-to-72-billion.

5. Weight loss spending has been growing between 2.6 percent and 4.1 percent while the US population growth is less than 1 percent annually (0.06 percent in 2018 according to WorldBank.org).

6. Although this might have been possible, at the time the seating structure was not in use and there is no evidence the facility was used to host sporting events at the time of Fermi's experiments. The University of Chicago discontinued their football program in 1939, and it did not return until 1963.

7. Bureau of Transportation Statistics actually shows a lower rate of production growth less than 1 percent, so it is rounded to 1 percent to make the math easier. "Annual U.S. Motor Vehicle Production and Domestic Sales," Bureau of Transportation, 2018, https://www.bts.gov/content/annual-us-motor-vehicle -production-and-factory-wholesale-sales.

8. In reality this goes back to October 1908 when Ford first started making the Model T. "Ford Motor Company unveils the Model T," History A+E Networks, https://www.history.com/this-day-in-history/ford-motor-company-unveils -the-model-t.

9. Erin Arvedlund, *Too Good to Be True: The Rise and Fall of Bernie Madoff* (New York: Penguin Publishing Group, 2010).

10. Harry Markopolos, "The World's Largest Hedge Fund Is a Fraud," November 7, 2005, submission to the SEC, https://www.sec.gov/news/studies/2009/oig-509/exhibit-0293.pdf.

11. Centers for Disease Control and Prevention, National Center for Health Statistics, updated August 2001.

12. F. J. Anscombe, "Graphs in Statistical Analysis," *American Statistician,* 27 (1)(1973): 17–21. doi:10.1080/00031305.1973.10478966.

13. Mars temperature averages –63 Celsius and ranges from –140C to +30C. Sarah Marquart, "It Is Colder Than Mars Right Now, but That's Not as Big of a Deal as You Think," *Science Alert,* February 1, 2019, https://www.science alert.com/it-s-colder-than-mars-isn-t-as-big-of-a-deal-as-you-think-it-is.

14. "How America Saves," Vanguard Investments, 2019, https://pressroom .vanguard.com/nonindexed/Research-How-America-Saves-2019-Report.pdf.

15. Readers might wonder if a data model can be easily created to separate written articles from fact or opinion when they might not see more online flagging of questionable news sources. The method we described is a basic approach and not technologically difficult, however, it was not until October 2016 that Google added fact-checks to their news queries. Facebook followed with a less robust approach in December 2016. Facebook contends they are trying to balance what users want to share and the need to fact-check their statements. Dave Lee, "Matter of Fact-checkers: Is Facebook Winning the Fake News War?" BBC News, April 2, 2019, https://www.bbc.com/news/technology-47779782.

16. IoT is an abbreviation for Internet of Things and refers to devices that traditionally are not part of the digital ecosystem. Many Wi-Fi-connected devices in your home such as your thermostat, washing machine, and irrigation system would all be considered IoT devices.

17. Computers store text one letter at a time with each letter assigned to one byte of data (eight bits of 1's and 0's). The name Shakespeare, for example, is stored as 11 bytes, one for each letter as 01010011 01101000 01100001 01101011 01100101 01110011 01110000 01100101 01100001 01110010 01100101. According to the Folger Shakespeare Library, the author wrote 884,647 words in his complete works; https://www.folger.edu/shakespeare-faq, accessed September 22, 2022.

18. Alvin Toffler, *Future Shock* (New York: Bantam Books, 1970).

19. Josephine B. Schmitt, Christina A. Debbelt, and Frank M. Schneider, "Too Much Information? Predictors of Information Overload in the Context of Online News Exposure, Information," *Communication & Society* (2018) 21:8, 1151–67, doi: 10.1080/1369118X.2017.1305427.

20. David P. Baker, Paul J. Eslinger, Martin Benavides, Ellen Peters, Nathan F. Dieckmann, and Juan Leon, "The Cognitive Impact of the Education

Revolution: A Possible Cause of the Flynn Effect on Population IQ," *Intelligence* 49 (2015): 144–58, doi:10.1016/j.intell.2015.01.003.

21. James R. Flynn, "Requiem for Nutrition as the Cause of IQ Gains: Raven's Gains in Britain 1938–2008," *Economics and Human Biology* 7 (March 2009): 18–27, doi:10.1016/j.ehb.2009.01.009.

22. J. E. Uscinski and J. M. Parent, *American Conspiracy Theories* (New York: Oxford University Press, 2014).

CHAPTER 4

1. Esquisse d'un tableau historique des progrès de l'esprit humain. Condorcet, 1795. The title can also be translated as "A Sketch of the Historical Progress of the Human Spirit." This was written while Condorcet was in hiding due to a warrant for his arrest in 1794 and was published after his death in 1795. The work presents several stages, epochs, of human enlightenment that shaped humankind through a broader understanding of scientific knowledge.

2. The first regularly published newspaper in the American colonies was the *Boston News-Letter* beginning in 1704. It was partly funded by the British government, and therefore it was licensed by the government and each edition was required to be reviewed and approved.

3. To Alexander Hamilton from George Washington, June 26, 1796, accessed March 3, 2021, https://founders.archives.gov/documents/Hamilton/01-20-02-0151.

4. Eric Burns, *Infamous Scribblers* (New York: Public Affairs Publishing, 2006).

5. For more on how social media crafts their content for the purpose of extending users' engagement, see the 2020 Netflix documentary "The Social Dilemma."

6. Elisa Shearer and Elizabeth Grieco, "Americans Are Wary of the Role Social Media Sites Play in Delivering the News," Pew Research Center, October 2, 2019, https://www.journalism.org/2019/10/02/americans-are-wary-of-the-role-social-media-sites-play-in-delivering-the-news.

7. In this study a fake news story was entirely made-up stories or facts. "Fake News, Filter Bubbles and Post-truth Are Other People's Problems," Ipsos MORI, September 6, 2018, https://www.ipsos.com/ipsos-mori/en-uk/fake-news-filter-bubbles-and-post-truth-are-other-peoples-problems.

8. Harriet Barber, "Finland's Secret Weapon in the Fight against Fake News: Its Kindergarten Children," *The Telegraph*, February 16, 2021, www.telegraph.co.uk/global-health/climate-and-people/finlands-secret-weapon-fight-against-fake-news-kindergarten/.

9. Facebook has long sought overall growth as its goal and has specific performance metrics such as the L6/7, which is the measure of the number of people who logged on in six of the previous seven days. Facebook has recently been increasing efforts to remove misinformation such as those related to vaccines, however, we could not find any meaningful metrics that measures the overall truthfulness in posts. Facebook's mission statement includes "give people the power to build community and bring the world closer together." In contrast, the *Washington Post*'s printed motto "Democracy Dies in Darkness" signals its intention is to provide accurate information in its reporting as a primary goal. The first of their seven goals, written in 1935 by Eugene Meyer, reads, "The first mission of a newspaper is to tell the truth as nearly as the truth may be ascertained."

10. Pengjie Gao, Chang Lee, and Dermot Murphy, "Financing Dies in Darkness? The Impact of Newspaper Closures on Public Finance," Hutchins Center for Fiscal & Monetary Policy at Brookings Institute, September 24, 2018, https://www.brookings.edu/research/financing-dies-indarkness-the-impact-of-newspaper-closures-onpublic-finance/.

11. In 2013 the US government officially acknowledged the existence of Area 51 following a Freedom of Information Act filing. It is located about 80 miles from Las Vegas and, due to its folklore, has become part of the tourism of Nevada. The highway that borders the facility was renamed by the state of Nevada as "Extraterrestrial Highway."

12. Sadly, this is a true story. In 2010 a Georgia man was jogging on the beach in Hilton Head, South Carolina, while listening to music. A small plane experienced engine trouble and had to make an emergency landing. Due to leaking oil the pilot could not adequately see the beach where he was landing and the jogging man was tragically killed; https://www.nbcnews.com/id/wbna35896336.

13. Bryan A. Strange, Andrew Duggins, William Penny, Raymond J. Dolan, and Karl J. Friston, "Information Theory, Novelty and Hippocampal Responses: Unpredicted or Unpredictable?" *Neural Networks* 18, issue 3 (2005):225–30, doi.org/10.1016/j.neunet.2004.12.004, https://www.sciencedirect.com/science/article/abs/pii/S0893608005000067.

14. Soroush Vosoughi, Deb Roy, and Sinan Aral, "The Spread of True and False News Online," *Science* 359, no. 6380 (March 9, 2018): 1146, https://doi.org/10.1126/science.aap9559.

15. Edson C. Tandoc Jr., Zheng Wei Lim, and Richard Ling, "Defining Fake News," *Digital Journalism* 6:2 (2018): 137–53, doi: 10.1080/21670811.2017.1360143.

16. The FTC complaint detailed that this practice of deceptively assuring users of privacy controls while at the same time providing access to third-parties to capture privacy data occurred over a seven-year period. United States

of America v. Facebook. Case 19-cv-2184 July 14, 2019, https://www.ftc.gov/
system/files/documents/cases/182_3109_facebook_complaint_filed_7-24-19
.pdf.

17. "FTC Charges Volkswagen Deceived Consumers with Its 'Clean Diesel'
Campaign," Federal Trade Commission press release, March 29, 2016, https://
www.ftc.gov/news-events/press-releases/2016/03/ftc-charges-volkswagen
-deceived-consumers-its-clean-diesel.

18. The eliminated FTC advertising guide for fallout shelters also once
included prohibition to give away shelters as part of a lottery and prevented
the use of scare tactics in advertisements "such as employing horror pictures
calculated to arouse unduly the emotions of prospective shelter buyers." It also
prevented the absolute guarantee of safety, which now seems like something
obviously no company could ever promise. We consider this as a sign that
we both live in safer times and consumers are less gullible to marketers over-
promising on the product's benefits;
https://www.google.com/books/edition/Code_of_Federal_Regulations/Aj6
eZ2vWiLQC?hl=en&gbpv=1&dq=229+Guides+for+advertising+fallout+
shelters&pg=PA17&printsec=frontcover.

19. Philip Reed and Nicole Arata, "What Is the Total Cost of Owning a
Car?" NerdWallet, June 27, 2019, https://www.nerdwallet.com/article/loans/
auto-loans/total-cost-owning-car.

20. According to the analysis, fuel efficiency due to driving habits varied
dramatically based on the type of vehicle. This amount of 38 percent was the
top end of the range and descended to 9 percent. The average was 31 percent.
Although the number of cars tested were small, the evidence shows there is
positive impact to less aggressive driving; Phillip Reed, "We Test the Tips,"
Edmunds.com, May 5, 2009, https://www.edmunds.com/fuel-economy/we
-test-the-tips.html.

21. "Fuel Economy in Cold Weather," FuelEconomy.Gov, US Department
of Energy, https://www.fueleconomy.gov/feg/coldweather.shtml.

22. We are also aware the economic depression was a factor in declining
tobacco sales.

23. The small print in the advertisement also read "Philip Morris do not
claim to cure irritation" and alluded to ingredients used by other manufactures
as the cause of irritation; *Saturday Evening Post*, Curtis Publishing Company,
October 16, 1937, 86.

24. Ernst L. Wynder, Evarts A. Graham, and Adele B. Croninger, "Experimental Production of Carcinoma with Cigarette Tar," *Cancer Research* 13,
1953.

25. The Frank Letter appeared in 448 newspapers in 258 cities and was
believed to reach more than 43 million readers. K. M. Cummings et al, "Failed
Promises of the Cigarette Industry and Its Effect on Consumer Misperceptions

about the Health Risks of Smoking," *Tobacco Control* 11 (March 2002):i110–i117, www.jstor.org/stable/20208011.

26. Dawn Connelly, "A History of Aspirin," *The Pharmaceutical Journal* (September 26, 2014), https://pharmaceutical-journal.com/article/info graphics/a-history-of-aspirin.

27. It was reported that Plavix advertisements in Dutch journals claimed a 26 percent reduction in complications compared with aspirin treatments. A. Algra, J. van Gijn, L. J. Kappelle, P. J. Koudstaal, J. Stam, and M. Vermeulen, Creatief cijferen met clopidogrel; overschatting van het preventief effect door de fabrikant [Creative mathematics with clopidogrel; exaggeration of the preventive effect by manufacturer], Ned Tijdschr Geneeskd, 143, 49 (December 1999). Dutch. PMID: 10608988. https://pubmed.ncbi.nlm.nih.gov/10608988.

28. G. J. Hankey and C. P. Warlow, "Treatment and Secondary Prevention of Stroke: Evidence, Costs, and Effects on Individuals and Populations," *Lancet* (1999) 354:1457–63, https://www.thelancet.com/journals/lancet/article/PIIS0140-6736(05)72318-7/fulltext.

29. The Plavix advertisement in March 2009 mentioned only stroke and heart attack and did not note the tremendous but less threatening benefits of PAD prevention. They wrote, "Plavix is proven to keep [blood] platelets from . . . forming clots . . . the leading cause of stroke. Plavix helps you stay protected." Aspirin also does the same.

30. The reason an iceberg floats is it is composed of frozen freshwater and is slightly more buoyant that the salt water it is displacing by its volume. "How Much of an Iceberg Is Below the Water," US Department of Homeland Security, United States Coast Guard, https://www.navcen.uscg.gov/?pageName=iip HowMuchOfAnIcebergIsBelowTheWater.

31. It is unlikely ancient Viking sailing boats would have suffered from sailing into an iceberg since they were traveling at slow speeds during daylight. Of course, famously, the steamship *Titanic* collided with an iceberg at night in 1912 while travelling at a speed of 22 knots.

32. Alison Hill notes the case fatality rate (CFR) is hotly debated among scientists, the media, and the general public. A better method of measuring CFR is to isolate cohorts rather than the simple ratio of current cumulative deaths to current cumulative cases in part due to data lags. The data show COVID-19 related deaths lag time of infection by about three weeks. Alison Hill, "The Math behind Epidemics," *Physics Today* 73, 11, 28 (2020), https://doi.org/10.1063/PT.3.4614.

33. "White House Tells Travelers from New York to Isolate as City Cases Soar," *New York Times*, March 24, 2020, retrieved June 22, 2021, https://www.nytimes.com/2020/03/24/nyregion/coronavirus-new-york-update.html.

34. Greg Rosalsky, "The Dark Side of the Recovery Revealed in Big Data," *Planet Money*, October 27, 2020, https://www.npr.org/sections/money/

2020/10/27/927842540/the-dark-side-of-the-recovery-revealed-in-big-data; reporting data from Opportunity Insights, https://opportunityinsights.org.

35. John Hopkins University & Medicine Coronavirus Resource Center, retrieved September 2022, https://coronavirus.jhu.edu/.

36. Andrew J Einstein, Leslee J. Shaw, Cole Hirschfeld, Michelle C. Williams, Todd C. Villines, Nathan Better, Joao V. Vitola et al, "International Impact of COVID-19 on the Diagnosis of Heart Disease," *Journal of the American College of Cardiology* 77, No. 2 (January 19, 2021): 173–85, https://doi.org/10.1016/j.jacc.2020.10.054.

CHAPTER 5

1. Carnival games have changed little over the decades, offering both simplicity of play and nostalgia. Fittingly the research showing most carnival games are unbeatable is also timeless as we reference this source from 1978. It seems we have known for a long time the games are chance, with little chance of winning. Donald A. Berry and Ronald R. Regal, "Probabilities of Winning a Certain Carnival Game," *The American Statistician* 32, no. 4 (1978):126–29. JSTOR, www.jstor.org/stable/2682938.

2. Jeff S. Barlett, "The Cost of Car Ownership over Time," *Consumer Reports*, April 2020.

3. Our formula and estimated repair costs are notably specific. This is simply due to the math and should not be confused with accuracy. Often, we believe when numbers do not round off to convenient whole numbers it lends to people overweighting their credibility. Our academic friends will also point out that creating a regression using only 50 data points (25 each for years 5 and 10) is not enough to make a solid estimate. However, the purpose of this example is to illustrate how to use easily obtained data to make reasonable everyday decisions.

4. Jose Miguel Abito and Yuval Salant, "The Effect of Product Misperception on Economic Outcomes: Evidence from the Extended Warranty Market," *Review of Economic Studies* 86, issue 6 (November 2019):2285–2318, https://doi.org/10.1093/restud/rdy045.

5. FTC website includes warranty requirements information and cautions customers to perform their own research on products reliability, which is not required to be written in the warranties; https://www.ftc.gov/enforcement/rules/rulemaking-regulatory-reform-proceedings/advertising-warranties-guarantees.

6. Rachel Premack, "How to Win the Lottery, According to a Romanian-born Mathematician Who Hacked the System, Won 14 times, and Retired on a Remote Tropical Island," *Business Insider*, March 21, 2019.

7. This is calculated as follows (18 black pockets / 37 total pockets)26-1.

8. Greeting Card Association Facts and Stats 2019.

9. "How Dangerous Is Lightning," NOAA (2019), https://www.weather.gov/safety/lightning-odds 2019.

10. "Skin Cancer Facts and Statistics," Skin Cancer Foundation, https://www.skincancer.org/skin-cancer-information/skin-cancer-facts 2020.

11. Kevin Williams, "Dynamic Airline Pricing and Seat Availability," *Cowles Foundation Discussion Paper* No. 2103R (May 26, 2020), https://ssrn.com/abstract=3611696 or http://dx.doi.org/10.2139/ssrn.3611696.

12. Ashley Kilroy, "Trip Cancellation Travel Insurance," *Forbes*, May 28, 2020, https://www.forbes.com/advisor/travel-insurance/trip-cancellation/.

13. "Consumer Sentinel Network Data Book 2019," *Consumer Sentinel Network Reports*, Federal Trade Commission, https://www.ftc.gov/enforcement/consumer-sentinel-network/reports.

14. Why pay more than the amount of the expected loss for insurance? The answer is there are additional costs in administering insurance and the insurance companies should expect to make a reasonable return on the risk they are investing into.

15. "NAPHIA's 2020 State of the Industry Report," May 26, 2020, https://naphia.org/about-the-industry/.

16. Technically the prize for economics is not a Noble Prize, but one administered by the Noble Foundation. The prize is officially known as the Sveriges Riksbank Prize in Economic Sciences in Memory of Alfred Nobel. The prize was not funded by Noble, but rather by a donation from Sweden's central bank, Sveriges Riksbank, in 1968.

CHAPTER 6

1. As of 2020 Texas had 157,343 oil wells, far more than any other state. The US total count of wells was 325,213. US Energy Information Administration, https://www.eia.gov/.

2. The Federal Reserve has consistently reported that about 26 percent of non-retired adults have zero retirement savings. Among those with savings only 36 percent believed their retirement savings were on track to meet their retirement goals; https://www.federalreserve.gov/publications/2021-economic-well-being-of-us-households-in-2020-retirement.htm.

3. We are assuming a simple annual compounding of an initial $10,000 investment as $10,000 × (1 + 9.9%)35.

4. John Allen Paulos, *Innumeracy, Mathematical Illiteracy and Its Consequences* (New York: Hill and Wang, 1988), 32–33.

5. US Mutual Funds Industry—Growth, Trends, COVID-19 Impact, and Forecasts, Research and Markets Inc., accessed October 20, 2021, https://www.researchandmarkets.com/reports/5394162/us-mutual-funds-industry-growth-trends-covid.

6. John Waggoner, "The 25 Best Mutual Funds of All Time," Kiplinger.com., October 21, 2019, https://www.kiplinger.com/slideshow/investing/t041-s001-the-25-best-mutual-funds-of-all-time/index.html.

7. Adjusted for stock splits, Microsoft traded for the equivalent of $0.11 per share and currently is valued at more than $300 per share. Annualized return over the 36-year period since 1985 is $(\$300/\$0.11)(1/36)-1$.

8. Xinge Zhao, "Exit Decisions in the US Mutual Fund Industry," *The Journal of Business* 78, no. 4 (2005):1365–1402. JSTOR, accessed February 9, 2021, www.jstor.org/stable/10.1086/430863.

9. "Why Worry about Survivorship Bias?" Dimensional, October 12, 2020, https://www.dimensional.com/us-en/insights/why-worry-about-survivorship-bias.

10. Referring to SEC rule 482 concerning advertising by an investment company.

11. Bill Gates' net worth (2021) is reported to be approximately $130.8 billion and Mark Zuckerberg's net worth is $130.6 billion (2021). Jeff Bezos' 2021 net worth is reported to be $190.7 billion, and he did graduate from Princeton with a degree in electrical engineering and computer science. These estimates are from *Forbes*.

12. Elka Torpey, "Education Pays 2020," US Bureau of Labor Statistics, June 2021, https://www.bls.gov/careeroutlook/2021/data-on-display/education-pays.htm.

13. "Homeowners vs Renters Statistics," iProperty Management, July 12, 2022, https://ipropertymanagement.com/research/renters-vs-homeowners-statistics#general-statistics.

14. John T. Cacioppo, Stephanie Cacioppo, Gian C. Gonzaga, Elizabeth L. Ogburn, and Tyler J. VanderWeele, "Marital Satisfaction and Break-ups Differ across On-line and Off-line Meeting Venues," *Proceedings of the National Academy of Sciences* 110, 25 (2013):10135–40, doi:10.1073/pnas.1222447110, https://www.pnas.org/doi/abs/10.1073/pnas.1222447110.

CHAPTER 7

1. CASE IH and John Deere are both working on experimental autonomous farm tractors. Joe Deaux, "Deere to Bring Fully Autonomous Tractor to Market This Year," Bloomberg, January 4, 2022, https://www.bloomberg.com/news/

articles/2022-01-04/deere-will-bring-fully-autonomous-tractor-to-market-this
-year.

2. An estimate for coffee shops' success or failure is by itself hard to measure. Most data suggest they survive the first few years and then close due to either the owners running out of enthusiasm or by declining capital.

3. "Connecting Small Business in the US," Deloitte 2018, https://www2.deloitte.com/content/dam/Deloitte/us/Documents/technology-media
-telecommunications/us-tmt-connected-small-businesses-Jan2018.pdf.

4. The committee resolved this by upgrading the ticket buyers to more expensive seats in the facility, although at a financial loss. Paul Keslo, "London 2012 Olympics: Lucky Few to Get 100m Final Tickets after Synchronized Swimming Was Overbooked by 10,000," *The Telegraph*, January 4, 2012, https://www.telegraph.co.uk/sport/olympics/8992490/London-2012
-Olympics-lucky-few-to-get-100m-final-tickets-after-synchronised-swimming
-was-overbooked-by-10000.html.

5. Eastman Kodak was summing up the displacement costs of employees that were being laid off. The error was for just one employee's severance expense. Jim Jelter, "Kodak Restates, Adds $9 Million to Loss," *MarketWatch*, November 9, 2005, https://www.marketwatch.com/story/kodak-restates
-earnings-adds-9-million-to-latest-loss.

6. In 2000 University of Hawaii professor Raymond Panko published a paper that reviewed several audits of spreadsheets going back 15 years with error rates ranging from 24 percent from a field study, 51 percent from a laboratory experiment with MBA students, and 90 percent for mechanical errors such as pointing a formula at the wrong cell. Interestingly a data dupe ensued as this report was repeated in the media. Somehow all those numbers appeared in an April 2003 *MarketWatch* article as "88 Percent of Spreadsheets Have Errors" and as recently as October 2019 when Oracle.com blogged "Almost 90 Percent of All Spreadsheets Have Errors." Both citied the same Panko 2000 study, and it is a mystery where 88 percent came from since it is nowhere in his paper.

7. Göran Ohlin. "The Population Concern," *Ambio* 21, no. 1 (1992): 6–9, JSTOR, accessed December 16, 2020, www.jstor.org/stable/4313877.

8. World population in 1981 was 4.54 billion and 5.5 billion by 1992. Population data by country is taken from Version 5 of Gapminder. This provides data by country from 1800 to 2100. The Gapminder data set is available online at http://www.gapminder.org/.

9. John von Neumann devised a method to create pseudo-random numbers using a middle-squares method and programmed the ENIAC, one of the first programmable computers.

10. "What the Cost of Airline Fuel Means to You," US Department of Transportation, September 2019, accessed December 18, 2020, https://

www.transportation.gov/administrations/assistant-secretary-research-and
-technology/what-cost-airline-fuel-means-you.

11. "More Cash Raised for Skybus," *Columbus Business First*, April 2,
2007, https://www.bizjournals.com/columbus/stories/2007/04/02/daily7.html.

12. London currently designs their flood plans for a 1-in-500-year event,
while the Netherlands plans for 1-in-1000-year event. In addition to the probabilities of such events given, the proximity to important infrastructure, such as
the cities, are a consideration. The financial losses and the economic disruption
weigh on the decision to plan for more unlikely events.

CHAPTER 8

1. Norse mythology attributed lightning and thunder to the god Thor, who
rode across the sky in a chariot. The scraping of the chariot's wheels caused
the light to flash, and the pounding of his mighty hammer created the thunder
sound we hear. Thor is a prominent figure in mythology and is the namesake
of "Thursday," as in Thors-day derived from the word Torsdag.

2. W. W. Patton, "Opening Students' Eyes: Visual Learning Theory in the
Socratic Classroom," *Law & Psychology Review* (1991).

3. Interestingly there was no actual research performed at Cambridge University as it claims. As the meme circulated, the reference to Cambridge was
apparently added to make it more believable and popular. However, unlike
being data duped, there is some truth to its claim about how the human mind
reads and processes words. Learning words has led to efficiency such that mis-
spelling only slows the reading processes slightly. For more on this topic see
the MRC Cognition and Brain Sciences Unit at real Cambridge University,
which created a website in response to the original meme; http://www.mrc-cbu
.cam.ac.uk/people/matt.davis/cmabridge/.

4. Shangri-la is fictional place from the novel *Lost Horizon* by James Hilton.
Since the novel's publication in 1933, the term "Shangri-la" has become synonymous with an idyllic place of perfection.

5. *Braveheart* took some historical liberty with the battle scene by combining the 1297 battle of Stirling Bridge, Wallace's most triumphant victory, with
the battle of Falkirk in 1298, which took place in a setting that most resembles
the scene in the movie.

6. The luckiest part of finding a four-leaf clover is finding it in the first place,
not any associated luck that follows. A couple of researchers in Switzerland
claimed to have studied more clovers than anyone and put the odds of finding
a four-left clover at more than 1 in 5,000. That is the equivalent to flipping a
coin and getting heads 12 times in a row. Lucky indeed.

7. Although psychologist Jean Piaget contributed tremendously to the understanding of the stages of child development, he concluded that when children reach the last one around age 12 and into adulthood, the Formal Operational Stage, then they increase this use of logic and deductive reasoning and think more scientifically about the world around them. Perhaps Piaget was also susceptible to his own confirmation bias as he somewhat believed other adults used rational thinking to understand and solve problems. For more see Harry Beilin, "Piaget's Enduring Contribution to Developmental Psychology," *Developmental Psychology* (1992).

8. In 1996 McDonald's Corporation's total advertising costs were $503 million. McDonald's Corporation Annual Report, 1996, page 39, https://archive.org/details/mcdonaldscorpannualreports/mcdonalds1996.

9. Paul Farhi, "The McMakeover of an Icon," *Washington Post*, May 8, 1996, https://www.washingtonpost.com/archive/business/1996/05/08/the-mcmakover-of-an-icon/.

10. The idea of loss-aversion was first studied by Kahneman and Tversky in 1979 and later by Kahneman in 2013. D. Kahneman and A. Tversky, "Prospect Theory: An Analysis of Decision under Risk," *Econometrica* (1979); David Kahneman, *Thinking, Fast and Slow* (New York: Farrar, Straus, and Giroux, 2013).

11. Irving L. Janis, *Groupthink: A Psychological Study of Foreign-Policy Decisions and Fiascoes* (Boston: Houghton, Mifflin, 1972; revised 1982).

12. Arthur M Schlesinger Jr., *A Thousand Days: John F. Kennedy in the White House* (Boston: Houghton Mifflin, 1965).

13. Hockey Hall of Fame, Legends of Hockey list, accessed February 9, 2021, https://www.hhof.com/html/legendsplayer.shtml.

14. Max Planck, *Scientific Autobiography and Other Papers* (London: Philosophical Library, 1950), 33–34.

15. R. S. Westman, *The Copernican Question: Prognostication, Skepticism, and Celestial Order* (United Kingdom: University of California Press, 2011).

16. Alan Cowell, "After 350 Years, Vatican Says Galileo Was Right: It Moves," *New York Times*, October 31, 1992.

17. First published in 1946, Proctor and Gamble print advertisements claimed, "No soap—no other product sold throughout America will clean as good as Tide." The claim may have been true but the advertisement was not backed by any data. However, they did offer a satisfaction guarantee refund if the customers did not agree they had the "cleanest wash in town"; https://ohiomemory.org/digital/collection/p267401coll36/id/22460/.

18. Staring at the Sun truthfully is a dangerous thing to do. We included it here to demonstrate there is a mixture of things we hear repeated. Some are true while others are not and we sometimes have difficulty knowing the difference based on our knowledge recall. Staring at the Sun for a prolong period

can cause solar retinopathy, which is recoverable. "How Long Does Solar Retinopathy Recovery Take?" NVISION, https://www.nvisioncenters.com/retinopathy/solar-retinopathy.

19. L. K. Fazio, D. G. Rand, and G. Pennycook, "Repetition Increases Perceived Truth Equally for Plausible and Implausible Statements," *Psychonomic Bulletin & Review* 26 (2019):1705–1710, https://doi.org/10.3758/s13423-019-01651-4.

20. Quentin J. Schultze and Randall L. Bytwerk, "Plausible Quotation and Reverse Credibility in Online Vernacular Communities," *ETC: A Review of General Semantics* 69, no. 2 (2012):216–234, www.jstor.org/stable/42579187.

CHAPTER 9

1. The total eligible voter population was larger than 45 million, although this was the number of votes cast in 1936. The actual poll population was 4.8 percent, just shy of 1 in 20 voters, and more accurately 1 in 20.7 voters. Other data were sourced from Peverill Squire, "Why the 1936 *Literary Digest* Poll Failed," *The Public Opinion Quarterly* 52, no. 1 (1988):125–33. JSTOR, www.jstor.org/stable/2749114.

2. Lily Rothman, "How One Man Used Opinion Polling to Change American Politics," *Time Magazine*, November 17, 2016.

3. In 1929, just prior to the crash of the stock market, telephone ownership had a peak of 41 percent of households. Telephone ownership declined during the Depression and by 1936 had dipped to 33 percent. Hannah Ritchie and Max Roser, "Technology Adoption in US Households," Our World in Data, accessed September 26, 2022, https://ourworldindata.org/technology-adoption.

4. Issac Newton, *Mathematical Principles of Natural Philosophy* (originally published 1687 and translated into English by Andrew Motte in 1729). This is the most common quote of Newton's Third Law of Motion, although the translated version by Motte reads "To every action there is always opposed an equal reaction; or the mutual action of two bodies upon each other are always equal, and directed to contrary parts."

5. Roger Fisher, William Ury, and Bruce Patton, *Getting to Yes: Negotiating Agreement without Giving In* (New York: Penguin, 2011).

6. We would be remiss not to mention that in 2007 a congressional investigation known as the Mitchell Report examined the extensive use of performance-enhancing drugs in baseball, including the period of the late 1990s. Several players acknowledged their use and the connection to their performance. Assuming Major League Baseball is sufficiently enforcing their ban on such substances, the data leaning toward recent years' increases in average runs and home runs gives rise to the claim these are the "best" years in baseball.

7. Jessica Yellin, "Single, Female and Desperate No More," *New York Times*, June 4, 2006, https://www.nytimes.com/2006/06/04/weekinreview/04yellin .html.

8. Holly Wyatt et al., "Long-Term Weight Loss and Breakfast in Subjects in the National Weight Control Registry," *Obesity Research* 10, no. 2 (February 2002).

9. Ahahad O'Connor, "Myths Surround Breakfast and Weight," *New York Times*, September 10, 2013, https://well.blogs.nytimes.com/2013/09/10/ myths-surround-breakfast-and-weight/.

10. Ibid.

11. An excellent resource is the Salem Witch Trials Documentary Archive and Transcription project hosted by the University of Virginia, http://salem.lib .virginia.edu/home.html.

12. One measure of the worst hurricane year might be total number of hurricanes and 2005 would top this list. With 15 named hurricanes, including 7 major hurricanes such as Dennis, Katrina, Wilma, and Rita, it was a devastating year. However, the worst year for deaths was the Great Hurricane of 1780, which reportedly killed more than 20,000 people. "Great Hurricane of 1780," History.com A+E Television Networks, April 12, 2019, https://www.history .com/topics/natural-disasters-and-environment/great-hurricane-of-1780.

CHAPTER 10

1. The internet has many roots and can date its early development back to the 1950s and 1960s along with the early development of computers. We chose 1983 since this was the time ARPANET (advanced research projects agency network) adopted the transmission protocol (TCP/IP) allowing computers to address others using their IP address.

2. Thomas Osborne, *Aspects of Enlightenment: Social Theory and the Ethics of Truth* (London: University College of London Press, 1998), 1–2.

3. Our apologies to credible snake oil salesmen. Genuine Chinese Snake Oil derived from the Chinese water snake has some merit in treating aching joint pain. The oil contained rich omega-3 acids and was used traditionally to treat inflammation. However, a scrupulous American entrepreneur named Clark Stanley tried to capitalize on the Chinese version of snake oil by concocting his own in 1897. The problem was it did not contain any snake oil or any other effective component. In 1917 federal investigators found Stanley's oil contained mineral oil, beef fat, red pepper, and turpentine. He was fined and put out of business without a fight. His contribution as a data(less) duper also contributed to our lexicon of "snake oil salesman," meaning someone who is deceptively selling a product. Lakshmi Gandhi, "A History of Snake

Oil Salesmen," *Codeswitch*, National Public Radio, August 26, 2013, https://
www.npr.org/sections/codeswitch/2013/08/26/215761377/a-history-of-snake
-oil-salesmen. For more information on omega-3 fatty acids and inflammatory
processes see: Philip C Calder, "Omega-3 Fatty Acids and Inflammatory Processes," *Nutrients* 2, no. 3 (2010):355–74, doi:10.3390/nu2030355.

4. Geoff Brumfiel, "The Life Cycle of a COVID-19 Vaccine Lie,"
National Public Radio, July 20, 2021, https://www.npr.org/sections/health
-shots/2021/07/20/1016912079/the-life-cycle-of-a-covid-19-vaccine-lie.

5. In June of 1812 Napoleon Bonaparte of France marched towards Russia. He had numbers on his side with an impressive army of 450,000, more
than double the estimated 200,000 Russian troops he expected to encounter.
However, as the push farther and farther into Russia towards Moscow continued through the winter months, Napoleon's supply chain grew weaker
and weaker, eventually forcing him to retreat. By the time he returned he
had fewer than 100,000 weary troops, and the event has been noted as
the turning point in his reign as emperor; https://www.history.com/news/
napoleons-disastrous-invasion-of-russia.

6. In early 2013 after years of development the Boeing 787, luxuriously
called the *Dreamliner*, was grounded. Not because it could not fly properly
or deliver on its many innovations in composite materials, fuel efficiency, or
systems. Rather, it was because of a relatively small box containing lithium-
ion batteries that were occasionally catching fire. Fire is a serious problem,
especially on a plane. Engineers, perhaps overfocused on the other complex
parts of the aircraft, made several assumptions about the data from quality
reviews and testing of the batteries from their suppliers, proving attention to
detail on small and large components when overlooked can equally disrupt a
project; https://www.nytimes.com/2014/12/02/business/report-on-boeing-787
-dreamliner-batteries-assigns-some-blame-for-flaws.html.

7. In 1999 the Mars Climate Orbiter (MCO) following a 10-month journey
from Earth plunged into the Martian landscape due to human math error. Two
separate teams responsible for parts of the orbiter built the system using different math scales—one metric and one imperial standard. The result was a
miscalculation of the acceleration force of the orbiter, measured in Newtons.
The data dupe might have been detected since the metric measure and standard
are more than an order of magnitude different (1 kg/meter/sec2 vs. 0.22-pound
force). The official reports on the matter note the obvious difference in math
and miscommunication. The miscommunication was attributed, as engineers
might, to a missing "specification," however it is more insightful to consider
how confirmation bias and the means in which humans make decisions also
contributed to its outcome. Robert Hotz, "Mars Probe Lost Due to Simple
Math Error," *Los Angeles Times*, October 1, 1999, https://www.latimes.com/
archives/la-xpm-1999-oct-01-mn-17288-story.html.

8. Jason H. Steffen, "Optimal Boarding Method for Airline Passengers," *Journal of Air Transport Management* 14, issue 3 (2008), https://doi.org/10.1016/j.jairtraman.2008.03.003.

9. "Forecast: Internet of Things—Endpoints and Associated Services Worldwide, 2017," Gartner, December 21, 2017, https://www.gartner.com/en/documents/3840665.

10. "Workforce of the Future: The Competing Forces Shaping 2030," PWC, 2018, https://www.pwc.com/gx/en/services/people-organisation/workforce-of-the-future/workforce-of-the-future-the-competing-forces-shaping-2030-pwc.pdf.

11. "Worldwide Quarterly Enterprise Infrastructure Tracker: Buyer and Cloud Deployment," International Data Corporation, September 2020, https://www.idc.com/tracker/showproductinfo.jsp?containerId=IDC_P31615.

12. Amazon may even do this more frequently, although 10 minutes has been reported.

Jerry Useem, "How Online Shopping Makes Suckers of Us All," *The Atlantic*, May 2017, https://www.theatlantic.com/magazine/archive/2017/05/how-online-shopping-makes-suckers-of-us-all/521448/.

Index

Active Measures Working Group, 27
Adams, John, 64–65
Adams, Samuel, 65, 68
advertising. *See* marketing
Age of Reason, 197
aggregation, 82–83
AI. *See* artificial intelligence
airline industry, 105–7, 148–50, 170,
    202, 220n6
Alexa, 21, 38
algorithms, 62
AMA. *See* American Medical
    Association
Amazon: Alexa, 21, 38; Amazon
    Web Services, 13; business at, 25,
    38; Facebook and, 21–22; Google
    and, 16, 23; Kindle, 23
AMD, 12
American Airlines, 170
American Conspiracy Theories
    (Uscinski and Parent), 61
American Gothic (painting), 54
American Institute of Public
    Opinion, 180
American Medical Association
    (AMA), 79–80
analytics: business, 3; data and,
    24–25, 133–35; data science and,

9–11, 19–20, 25–26; deception
    with, 131–33; decisions with,
    135–40; decision-theory in,
    117–18; Google Analytics,
    134–35; Industrial Revolution
    compared to, 24–25; labor in,
    153; in modernity, 13–16; over-
    fitting and, 53–54; revolution,
    195, 197–98; simulations from,
    145–50; spreadsheets from,
    140–45, *141*, 153; statistics and,
    150–53; technology for, 4, 11–13;
    traffic, 17–18
animals, 160
Anscombe, Francis, 43–44
Anscombe's Quartet, 43–44, *44*, 47
Apollo space mission, 12
Apple, 16–17
approximations. *See* estimation
Area 51, 209n11
Aristarchus of Samos, 175
artificial intelligence (AI), 17–18,
    25
Art of War (Sun Tzu), 59–60
Asch, Solomon, 168–69
assumptions, 8, 198
astronomy, 175, 196–97
Australia, 22, 152

in, 109–12; information in,
2–3, 59–61; knowledge in,
25; misinformation in, 59–60;
psychology of, 20–24
Mona Lisa (painting), 54
money, 124–25
Monte Carlo Casino, 101
Monte Carlo simulations, 121–22,
146–50, 153
Morgan Stanley, 148–49
mutual funds, 126–27

NAPHIA. *See* North American Pet
Health Insurance Association
Napoleon Bonaparte, 220n5
National Association of Stock Car
Racing (NASCAR), 31
National Treasure (film), 30–31
natural disasters, 104, 108–9, 151–
52, 191–92, 216n12, 219n12
natural language processing (NLP),
20–21
Nazis, 176–77
Nernst, Walther, 174
Netflix, 23
Newcomb, Simon, 42
New Deal, 181, 188–89
news: COVID-19 in, 86; culture, 63–
64, 74; data and, *74*; fake, 64–66;
history of, 208n2; malarkey in,
92; *Newsweek* magazine, 185–87;
sensationalism in, 72–73; from
traditional media, 66–69; trust in,
69–70; truth in, 70–74
newspapers. *See* journalism; media
Newsweek magazine, 185–87
Newton, Isaac, 182–83, 218n4
Nightingale, Florence, 9–11
Nightingale Graphs, 11
NLP. *See* natural language
processing
Norse mythology, 216n1

North American Pet Health
Insurance Association (NAPHIA),
112
Notes on Nursing (Nightingale),
10–11
Novum Organum (Bacon), 196–97
nuclear reactors, 31–32

O'Donnell, Rosie, 185
Ohlin, Göran, 143–44
online dating, 129–30
open minds, 177–78, 191–93
"Operator" (song), 21
opinions: American Institute of
Public Opinion, 180; bias and,
53–54; facts and, 5, 62; open
minds and, 191–93; opinion data,
55–59
Osborne, Thomas, 197
Ottoman Empire, 10
outliers, 44–47, 92, 140, *141*, 188,
191
over-fitting, 53–54

PAD. *See* peripheral arterial disease
Panko, Raymond, 215n6
Parent, Joseph M., 61
Patil, D. J., 9
patterns: Benford's Law and, 40–42;
in data, 30–31; decisions from,
155–57; minds and, 157–59
Patton, Bruce, 183
Paulos, John Allen, 84, 125–26
people, 179–85, 193
peripheral arterial disease (PAD),
81–83
personalization-privacy paradox,
23–24, 26
pet insurance, 112–13
Pet Insurance Review, 112–13
Philip Morris, 78–79, 210n23
physics, 182–83, 218n4

About the Authors

Derek W. Gibson has spent most of his career providing business analytics in financial services guiding business strategy and decision making, primarily with Wachovia and Wells Fargo Bank. He has built expertise from back-room operations to the front office, with experience in banking, retirement, trust operations, finance, marketing, and client experience analytics.

His analytics specialty is bringing together people, sharing his knowledge of the business, and applying data science to discover opportunities. He is a leader of client analysis with a focus on delivering client insights influencing the business by using advanced statistics to discover trends and patterns of customer behavior.

Derek also serves on the Wake Forest University MS Business Analytics advisory board and has taught as an adjunct professor of data analytics and business models. He is a frequent speaker to the next generation of data scientists on the trends and needs in the business world.

Jeffrey D. Camm is the Inmar Presidential Chair in Analytics and the academic director of the Center for Analytics Impact at the Wake Forest University School of Business. A firm believer in practicing what he preaches, he has consulted for numerous corporations, including Procter and Gamble, Owens Corning, GE, Duke Energy, Tyco, Ace Hardware, Starbucks, and Kroger. In 2016, he received the Kimball Medal for service to the operations research profession, and in 2017 he was named an INFORMS Fellow. He is coauthor of ten books in statistics, management science, and analytics.